MW00846112

MODELING AND APPROXIMATION IN HEAT TRANSFER

Engineers face many challenges in design and research involving complex thermal systems. In practice, modeling and associated synthesis very often precede sophisticated analysis or even the need for precise results. *Modeling and Approximation in Heat Transfer* describes methods for reaching engineering models of physical systems by approximating their characteristics and behavior. This textbook provides a systematic discussion of how modeling and associated synthesis can be carried out by identifying key first-order effects, by making order of magnitude estimates, and by making bounding calculations. This book aims to help students develop greater facility in breaking down complex engineering systems into simplified thermal models that allow essential features of their performance to be assessed and modified. Topics include steady and unsteady heat conduction, convective heat transfer, heat exchangers, and thermal radiation. Many worked examples are included.

Leon R. Glicksman is Professor of Building Technology and Mechanical Engineering. He founded and served as the head of the Building Technology Program for 25 years. He has worked on fluidized beds, glass forming, natural ventilation, and sustainable design for developing countries. He is a Fellow of the American Society of Mechanical Engineers (ASME) and the American Society of Heating, Refrigerating, and Air-Conditioning Engineers (ASHRAE). He was awarded the 1969 ASME Robert T. Knapp Award, the 1970 ASME Melville Medal, and the 2008 ASME Heat Transfer Memorial Award.

John H. Lienhard V has been a Professor of Mechanical Engineering at MIT for about 30 years, working in heat transfer, thermodynamics, fluid flow, and desalination. Lienhard is a Fellow of the American Society of Mechanical Engineers (ASME) and a registered professional engineer. His awards include the 1988 National Science Foundation Presidential Young Investigator Award, the 2012 ASME Technical Communities Globalization Medal, the 2015 ASME Heat Transfer Memorial Award, and several teaching awards.

Modeling and Approximation in Heat Transfer

Leon R. Glicksman
Massachusetts Institute of Technology

John H. Lienhard V
Massachusetts Institute of Technology

CAMBRIDGE
UNIVERSITY PRESS

CAMBRIDGE
UNIVERSITY PRESS

One Liberty Plaza, New York, NY 10006, USA

Cambridge University Press is part of the University of Cambridge.

It furthers the University's mission by disseminating knowledge in the pursuit of education, learning, and research at the highest international levels of excellence.

www.cambridge.org
Information on this title: www.cambridge.org/9781107012172

First published 2016

Printed in the United Kingdom by Clays, St Ives plc

A catalog record for this publication is available from the British Library.

Library of Congress Cataloging in Publication Data
Glicksman, Leon R., 1938 author.
Modeling and approximation in heat transfer / John Lienhard, Massachusetts Institute of Technology, Leon Glicksman, Massachusetts Institute of Technology.
   pages   cm
Includes bibliographical references.
1. Heat – Transmission – Mathematical models.   2. Heat engineering – Mathematics.   I. Lienhard, John H., 1961 author.   II. Title.
TJ260.L4453   2016
621.402′2011–dc23   2015032009

ISBN 978-1-107-01217-2 Hardback

Cover photo: View of the space shuttle Atlantis, reentering the Earth's atmosphere, photographed by the International Space Station (courtesy NASA). See homework problem 3.12 for approximate model of reentry heat transfer.

To

Judith Kidder Glicksman and Theresa Kavanaugh

Contents

Preface

Many excellent textbooks on thermal science have been written, often including detailed physical theory and mathematical analysis. Yet, most of these texts lack any systematic discussion of how modeling and associated synthesis can be carried out – in engineering practice, these steps very often precede mathematical analysis or the need for precise results. Some may argue that these initial stages of problem formulation are a skill that cannot be taught formally and that can be acquired only through long experience with engineering problems. The present authors disagree with that premise.

During the past 15 years, we have developed a course on modeling and approximation of thermal systems, which we have offered to graduate-level engineering students at MIT. This text has evolved from our experience in teaching that course. The course owes much to the traditional structure of the MIT doctoral qualifying exams as well – which many outside MIT are surprised to learn focus primarily on assessing a candidate's ability to think in physical terms, rather than on his or her ability to do fancy calculations! We have found that after completing this course students have greater facility in breaking down complex engineering systems into simplified thermal models that allow essential features of the system's performance to be assessed and modified.

Learning this material requires working through problems that apply these ideas in an engineering context. We have included end-of-chapter problems of this nature. Most are open-ended without a unique "correct" solution and some appear to be very complicated. We have found that students who take the time to work through these exercises come away with a good understanding of the material. In our own teaching, we generally assign the problems to students to work ahead of class meetings, and we devote a substantial amount of the classroom time to discussing how to approach the problem, the possible solutions, and the various factors and complications that they contain. This discussion involves the entire class in a question-and-answer format.

The minimum background assumed of readers of this book is a one-semester undergraduate level subject in heat transfer, along with its normal prerequisite subjects in thermodynamics and fluid mechanics. Prior study of heat transfer at the graduate level will certainly be helpful as well. In addition, students who have had the opportunity to do some work on the design or experimental characterization of thermal systems are also likely to have experienced the challenges that this book undertakes to address.

Introduction to Modeling Real-World Problems

1.1 Introduction: Goals of this Book

In many situations faced by the product designer or researcher, the initial step – how to formulate the problem – is unclear, and the presence of multiple thermal phenomena complicate this initial step further. The central challenge is often to make a simplified model of the system or process and to obtain some first-order approximations for the magnitude of the key variables. This textbook presents an introduction to methods of thermal modeling and associated approximation tools. A broad rang of techniques is discussed. Some will give only order of magnitude estimates, while others may provide very close approximations, often without laborious calculations. We will generally approach these tools and methods as means of evaluating new, initially complicated heat transfer problems, identifying the governing physical phenomena and establishing their magnitude. Our hope is to provide readers with a much greater facility to deal with these challenges.

In this chapter, we review the engineering approach for defining a problem and exploring possible solutions. Several broad tools are considered, ranging from examination of the very general context of the problem to specific steps used to begin an initial estimate of the performance. The application to complex real-world problems is contrasted with the more controlled approach used in simplified academic exercises.

The first step in proper modeling is to make a careful definition of the goals of the project. The important factors to be considered in identifying goals are discussed next. The balance of the first chapter is devoted to an introduction to modeling techniques that can be employed in the start-up phase of the new concept or design. These same techniques are illustrated in the succeeding chapters, where we will develop modeling approximations for a number of different heat transfer processes.

1.2 The Art of Engineering: What Textbooks Don't Cover

Traditional textbooks in heat transfer and other thermal sciences concentrate on the development of fundamental principles and their corollaries, usually through the analysis of well-defined situations with relatively simple geometries. In many cases, a closed-form solution results. Emphasis is placed on obtaining very accurate solutions to very well defined problems. Although this may be a satisfactory way to initiate students into a new field, it leaves many ill prepared to deal with challenges that arise in engineering practice.

In the design of new processes or innovative techniques to improve present systems, problems are seldom well defined. It is often unclear what analysis is required. In the initial design phase, thermal performance must be evaluated in concert with many other considerations such as size, cost, weight, and other indices related to (for example) electronics or kinematic performance. Usually, many different designs are being considered simultaneously. What are needed at this stage are modeling techniques and tools that allow the engineer to do a first-order estimate of thermal performance that can be used in the initial design stage.

Consider the opposite approach: the temptation is to immediately begin a detailed comprehensive thermal analysis of specific designs, usually with a major component of the analysis being carried out by computer. The analysis is comprehensive enough to include all possible parameters that may come into play. Alternatively, an engineer working on improvements to a complex existing system may be unable to define and carry out a complex analysis, and may instead react by carrying out a detailed cycle of experimental testing in which multiple parameters are varied across a wide range.

The problem with either the detailed analysis or the experimental approach is twofold: the engineer does not initially concentrate on the most important aspects of the problem – understanding what is really happening – and in the rush to develop complicated computer solutions or experiments no time is left to separate the forest from the trees. In many cases, this leads to very sophisticated analyses or detailed experimental measurements of the wrong problem.

EXAMPLE

The need for proper synthesis and modeling is best illustrated by a practical example. In the following, we discuss how simple models can help guide the engineer in defining the problem and exploring possible solutions.

> **Example 1.1** *Office copying machines* Photocopiers use static charge to attach fine black particles, the toner, to a drum and then to blank paper. In large copying machines, the toner is fused to the paper by heating it while simultaneously applying pressure. The paper, with the toner attached, is passed over a heated roll called a fuser roll (see Fig. 1.1). A second roll presses the paper tightly against the fuser roll. To avoid delays in warming up the system, the fuser roll is kept at an elevated temperature when the machine is idling between jobs. The heat lost from the fuser roll results in substantial standby energy consumption.
>
> The goal is to design a copier that is very energy efficient while still offering quick startup. The desire for fast on-demand performance from an office copying

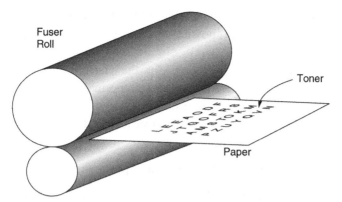

Figure 1.1. Fuser roll configuration for copying machines.

machine is in conflict with the desire to make the office equipment more energy efficient. In particular, current copiers implement a fast startup feature by keeping the fuser roll warm at all times, thereby using substantial energy in standby mode.

An innovative new design would maintain rapid response while doing away with most of the heat loss. There are several possible design strategies. One approach is better insulation of the fuser roll to minimize heat loss while it is idling at elevated temperature. Another is to find a means to heat the fuser roll rapidly when copies are needed. In this case, the fuser roll is kept at low temperatures during idling, reducing the heat lost. Still another possibility is the development of a new technique to fuse the print on the paper without the need for a fuser roll.

Before embarking on any of these design innovations, it would be well to verify that the heat lost from the fuser roll is the major source of energy consumption during the idling period. This involves approximating the heat lost from the surface of the fuser roll by convection and radiation, as well as estimating the conduction heat loss through the bearings and supports of the roll. A simple measurement of the energy consumption during idling may be called for to check those estimates.

The estimation of the energy loss from the fuser roll will also identify the most important sources of heat loss. This can point to the most effective ways to reduce that loss while the fuser is idling at high temperature. For example, if infrared radiation is the largest source of heat transfer from the fuser roll, the use of radiation shields around the fuser and the use of a low-emissivity coating on the fuser could be effective means to improve the copier's energy efficiency.

To explore the second strategy, rapid fuser heat-up, a completely different set of issues must be investigated. These issues include possible designs of a fuser with a low thermal capacitance (so-called thermal inertia) or a more rapid way of heating the surface of the fuser in contact with the paper.

A technology that omits the fuser would include consideration of new ways to heat the paper and print, possibly without direct contact, such as infrared sources.

These alternative design options for the energy-efficient copier can be explored effectively by the use of very simple thermal models. Such simple thermal models are appropriate as part of an overall design exploration that must also include possible new geometries and various materials of construction.

1.3 The Engineering Approach

The focus of this text is on thermal modeling. However, it is important to think of modeling in the context of the overall design project or desired problem solution. To begin a design process, several broad rules should be considered, as we outline next. That is not to say each project is best handled by following these steps by rote. The nature of the creative process requires considerable flexibility in the thought process as well.

The general steps to formulating a useful model follow this pattern:

- Identify what is the overall objective. What are the performance criteria in terms of energy efficiency, weight, and speed of operation, for example? What are the costs, time frame, and resources available to achieve these? If this is an improvement of an existing process, what are the key bottlenecks or shortcomings of that process?
- What is the initial set of likely new systems that can possibly meet the performance and other, desired criteria? For existing processes, what new concepts might help overcome the bottlenecks? For each of these prospective systems, an initial estimate of performance should be obtained. The focus here will be on the estimates of performance, but it should be understood that the other criteria must also be examined. In some cases, criteria unrelated to physical performance may be the overriding concern.
- For the system selected, an approximate definition should be stated. Again, what are the most important performance characteristics and system constraints? Attention should be focused on key elements of those performance criteria. A typical system configuration is assumed; alternatively, desired performance goals are set and the system configuration to meet those goals is found.
- Approximations are made about the system. Here the goal is to simplify the initial modeling and focus on the controlling parameters. The process might include developing assumptions that simplify the modeling. The assumptions will need to be justified, for example, by order of magnitude calculations.
- The performance of the system is estimated keeping in mind the level of detail necessary at this stage. The results are used to compare various design concepts and to eliminate those that are clearly inferior or cannot reach the desired goals. It is also important at this stage to ensure that the approximations, system definitions, and performance estimates look reasonable. That is, does the model give

believable results? If it does not, the approximations, definitions, and modeling steps need to be repeated. One simple check is to use the same modeling steps for a related, simplified system in which the performance or solution is known.

- The performance of the new system must be considered in light of existing devices and desired performance improvements. In some cases, it will be necessary to make a fresh start, identifying other candidate system designs.

Some engineering textbooks formulate problem solving into a rigid sequence of steps that resemble the latter steps in the preceding list. The problem is usually narrow and well defined and, in some examples, the context of the problem is not even stated. It is difficult to judge whether a more complete solution is required. The unspoken assumption is that a very thorough and complete solution of the narrowly defined task is the only "correct answer." Finally, although the steps outlined were given in the context of a design challenge, an analogous approach applies to the initial steps in the research of a new area.

1.4 Too Many Significant Figures

In many introductory scientific and engineering subjects, students are expected to come up with answers that in many cases represent an exact solution. If one carefully looks at a number of books in a field such as fluid mechanics or heat transfer, the same situations and boundary conditions are always considered in the derivations. These cases represent important limiting solutions and are valuable for study. The same examples appear repeatedly because they are the only ones for which an exact elegant closed-form solution is available. This can mislead readers into the expectation that all real problems will yield similarly elegant solutions and that such a level of precision can routinely be obtained. Unfortunately, when faced with a real-world problem, the situation is often so complicated that an exact solution is out of the question. On the other hand, a precise solution is usually not required. For the design example in the previous section, preliminary comparisons sometimes require only an order of magnitude approximation.

Are extended computer solutions with jazzy color graphics the order of the day? Enlightened managers and customers will shy away from such computer overkill, especially when they see the manpower requirements or the consulting bill for such an effort. In particular, when the system is too narrowly defined or does not consider all of the most relevant factors at the beginning of a design process, a solution with more than one significant figure is generally useless. The engineer needs to keep in mind what level of accuracy or approximation is needed for the question at hand: What is needed to make a design decision; what factors are most important; what are the real constraints; and what is controlling the physical process?

The administration in a Caribbean island was considering a solar desalination system. They wanted an accurate estimate of the flat plate thermal collector size that is needed to produce the desired output. Accuracy was needed because the cost of such a system was very high and overcapacity would be costly. Because the desired goal was increased freshwater supply, before doing an accurate estimate the set of possible solutions was broadened. It was found that the annual rainfall was high; a modest sized horizontal surface for rainfall collection gave the same results as the much larger solar collector (whose minimum size could be quickly estimated).

1.5 Property Values

In many situations, there is a large degree of uncertainty in thermophysical property values. Again, it doesn't make sense to carry out a numerical or analytical solution to three or four significant figures when the controlling property values may only be known within $\pm 25\%$ This limitation is frequently overlooked when engineers or scientists produce "very precise solutions."

Handbooks and textbooks contain a variety of tables giving thermophysical properties. What is often omitted is the range of *uncertainty* of the reported data. For even such a mundane and commonplace material as steel, the thermal conductivity can vary by as much as a factor of four as the carbon and chromium content of this steel is varied between pure iron and stainless steel. Thermal radiation properties of a shiny metal surface can vary over time by a factor of five or ten as the surface is oxidized or becomes dirty! Not only is a high precision calculation uncalled-for in such a case, but it also may be totally misleading if the properties are poorly understood.

As an example, manufacturers of polyurethane closed-cell foam insulation are concerned about predicting the thermal conductivity as a function of time. Foam insulation is made with a high molecular weight, low-conductivity gas contained in closed cells. When newly made, these foams have a higher resistance to heat transfer than almost any other conventional insulation of comparable thickness. Over time, air diffuses into the foam interior, mixes with the lower conductivity blowing gas, raising the effective conductivity of the foam by 30% or more. To find the rate at which the air concentration increases within the foam, one needs only to solve a one-dimensional diffusion equation – Fick's equation for mass transfer of the intruding air and original gas. The solutions exactly parallel the well-known solutions for one-dimensional transient heat conduction in a homogeneous body. Unfortunately, the diffusion coefficients of the various gases through the foam matrix are poorly known. Some measurements of the diffusion coefficient disagree by more than an order of magnitude. The main resistance to diffusion is the series of solid cell walls. In some foams, the solid polyurethane is not uniformly distributed, the center of the cell walls are much thinner than the edges. What was initially conceived as a project that could be carried out in a matter of weeks, ended up as a multiyear project to measure the diffusion coefficients and the foam morphology accurately!

1.6 Introduction to Modeling Tools and Techniques

One of the most challenging questions posed by a new design, research project, or application of an existing system is "How do we get started"? At this stage the engineer is faced with the interplay of ideas, constraints, and performance goals. It is important to determine quickly if the approach, design, or physical model is on the right track. This usually requires a quick estimate at the beginning. In many cases, this may be sufficient to rank order competing proposals or to identify the correct direction for further inquiry. The estimate may relate to the thermal efficiency, the maximum capacity for heating and cooling, or the safe temperature levels of operation.

In the starting phases of the project it is necessary to define the goal. Is it a new or improved project? What are the technical performance goals, and how important are these relative to reliability or cost? The level of accuracy needed for the performance prediction should be established. If the new concept is intended to double a performance parameter then initial estimates of the performance that have, say ± 10% uncertainty, may be more than adequate for initial evaluations. If cost is the main consideration, and it is not known with high precision (which is the usual case), then high precision in technical performance predictions may not carry much weight in the overall evaluation of the concept.

In other cases in which a high degree of accuracy is required for predicting the physical behavior, an initial, more approximate approach may still be the proper first step.

The watchword of the opening study is to keep things as simple as possible, sometimes referred to, humorously, as the KISS rule: "Keep it simple, stupid!" This approach has several advantages:

- It forces the modeler to concentrate on understanding the controlling physics without the possible confusion of elaborate analytical or numerical results.
- It can be carried out quickly – a virtue when there is a short deadline.
- It provides a quick check on the feasibility of the proposed approach.
- It is far easier to explain to managers who may not be technically adept.
- It establishes an initial credibility for your efforts and sometimes helps to limit unnecessary effort or going down a blind alley.

The authors have evaluated many projects that have been carried out in elaborate detail when a far simpler approach would demonstrate that the concept just won't work or is far inferior to other options. The goal of this text is to introduce such an approach, with the use of modeling and simplified tools as the proper starting point.

1.7 Modeling Techniques

A number of simple techniques can be employed in the startup phase of a new concept or design. There is no set order or pattern in which they should be applied and some may not be useful for a particular case. They are introduced in the following paragraphs and discussed in detail throughout the text.

Order of Magnitude Estimates

Order of magnitude estimates are very useful to identify which physical mechanisms are controlling and, equally important, which mechanisms can be neglected as

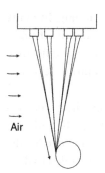

Figure 1.2. Glass fiber spinning.

second-order effects. Based on the approximate size, temperature, and other parameters, it is usually possible to make an order of magnitude estimate. How large must
a heat exchanger be to provide a heat transfer rate of a certain magnitude? For a
given power level and surroundings, how high a temperature will a body achieve?

Energy Balance

In thermal problems, an approximate energy balance will help identify important
parameters. For example, if the heat generated within a body at steady state is
100 W and the heat transferred from it to the surrounding air by convection can
be estimated to be about 90 W then conduction to adjoining solid bodies and radiation to the surroundings are second-order effects and may reasonably be omitted
from initial consideration. If the convective heat transfer is estimated to be no more
than 10 W, then it can be neglected and attention should be focused on radiation
and/or conduction. For a proposed new design with a heat generation of 100 W, if
the levels of conduction, convection, and radiation are all at most 10 W or less at the
proposed design temperature, the design is most probably a dead end or in need of
major revisions: there's no need to study it in greater detail without changing it first.

EXAMPLE

The use of modeling and approximations can sometimes be thought of as part of the
"art" of engineering. To master this art requires practice with a variety of situations.
Readers are advised to read the examples and to attempt their own modeling of
the situations presented. In most cases, there is not a unique "correct" solution. The
usefulness of the process depends on the ingenuity and creativity of the individual.
In all cases, it is good to keep in mind that the progression of approaches should start
with the simplest model first.

Example 1.2 *Glass fiber spinning* Continuous glass fibers are used for reinforc-
ing polymers in structures such as boats and car bodies. They are also used in
textiles. The glass fibers are formed from molten glass (see Fig. 1.2). The glass
is heated to about 1000°C, at which temperature it is a liquid with a viscosity
similar to that of heavy motor oil or syrup. It is delivered to a platinum vessel

that has a thousand or more small nozzles or tips on the bottom. The glass flows out of the tips and forms continuous strands or fibers that are pulled vertically downward. The fibers are cooled by a horizontal flow of ambient air. They are gathered at the bottom and wound around a rotating wheel. The wheel applies tension to the glass and attenuates it down to its final diameter as the glass is cooled.

An important design parameter of the process is the amount of ambient temperature air that must be blown horizontally across the glass to achieve a temperature drop of the fibers of 500°C.

The first estimate of the airflow can be obtained from an energy balance. For a control volume around all of the fibers in steady state, neglecting radiation from the glass to the surroundings,

$$(\dot{m}c_p\Delta T)_{\text{glass}} = (\dot{m}c_p\Delta T)_{\text{air}} \tag{1.1}$$

where $\dot{m}$ is the mass flow rate, c_p is the specific heat capacity at constant pressure, and ΔT is the change in temperature. The entire energy change of the glass can be considered sensible heat; there is not an abrupt phase change as the glass cools. The mass flow of air depends on the air temperature rise as it passes over all of the fibers.

Furthermore, for all of the glass fibers to cool evenly, each fiber, nearest the air flow entrance and those furthest away, must experience a similar temperature level of the air so that for convective cooling

$$\dot{m}\,c_p\Delta T_{\text{glass}} = hA_{\text{fiber}}(T_{\text{glass}} - T_{\text{air}}) \tag{1.2}$$

where h is the convective heat transfer coefficient and A_{fiber} is the surface area of all the fibers. It can be shown that radiation is a negligible source of cooling once the glass fiber has been reduced in size.

For the cooling to be approximately the same for all fibers, the change in the air temperature must be much smaller than the glass temperature change,

$$\Delta T_{\text{air}} \approx \frac{1}{10}\Delta T_{\text{glass}} \tag{1.3}$$

Combining this with the energy balance, an initial estimate of the air flow can be made. The next step is to ensure a uniform distribution of air over all of the fibers. Near the bottom of the platinum vessel the air is heated by convection from the vessel surface and the platinum nozzles. Our estimate did not take this heating into account so it must be treated as a first, lower limit estimate, for the proper air flow rate.

Maximum/Minimum Bounds

Although systems with complicated geometries usually require intense efforts to obtain close predictions of their performance, it may be sufficient to bracket the correct answer within upper and lower bounds or limiting solutions. This can be

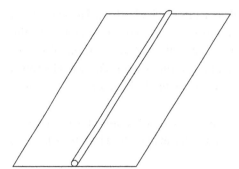

Figure 1.3. Wire-heated plate.

accomplished by making several simplifications to the geometry or assumptions about the process or system that will result in answers that can be easily shown to either exceed the correct results or provide a lower limit. The upper and lower limits may provide sufficient bounds for the purposes of an initial evaluation. In some cases, upper and lower bounds can be found that are very close to one another, resulting in a good estimate of the exact solution.

Example 1.3 *Microprocessor chip* Consider a microprocessor chip. Its steady-state energy consumption is 10 W and it is to be cooled by fins with a total area of 0.01 m^2. The designers are concerned about the maximum chip temperature relative to the surrounding air temperature. If the cooling relies on natural convection, then the minimum possible temperature difference between the surface of the chip and the surrounding ambient air occurs when the fin surface temperature everywhere is equal to the chip temperature,

$$\Delta T_{\min} = \frac{Q}{hA_{\text{total}}} \tag{1.4}$$

where Q is the heat flow out of the chip. Taking an upper limit on h for natural convection as 25 W/m^{2}K, then the minimum ΔT between the air and chip with natural convection is 10/[(0.01) (25)] or 40°C.

An extension of this example can be used to establish upper bounds, as well. To provide a heated surface for temperature control of an experiment, a heater wire is attached to one side of a vertical thin flat plate (Fig. 1.3). The wire is 10 cm in length with a diameter of 2 mm. The overall heating of the wire at steady state is 2 W. The plate dimensions are 10 cm by 5 cm, and it is 5 mm thick. The wire and plate are surrounded by ambient air. An upper bound on the wire temperature will determine whether the wire will soften or melt the plate adjacent to it. The lower limit of wire temperature is obtained by assuming the plate is highly conductive and that the wire is in good contact with it, similar to the preceding microprocessor example.

An upper bound on the wire temperature is obtained by assuming that the plate is made of very low conductivity material (a plastic, for example) or by assuming that a large contact resistance exists between the wire and the plate. In this extreme, negligible heat is transferred between the wire and the plate; and at steady state, all of the energy dissipated in the wire must be transferred from the surface of the wire to the surrounding air. Thus,

$$Q = I^2 R = hA_{\text{wire}}(T_{\text{wire}} - T_{\text{air}}) \tag{1.5}$$

where I is the electric current and R is the electrical resistance of the wire. Considering natural convection with a value of h between 5 and 25 W/m^2 K, an estimate of the wire temperature becomes

$$T_{\text{wire}} - T_{\text{air}} \approx \frac{2}{5(0.1 \times 2 \times 10^{-3})} = 2000°\text{C} \tag{1.6}$$

The very high temperature indicates that there will be severe problems in this limit. It also suggests that a better estimate of the wire temperature should include radiation from the wire surface, which requires an estimate of the wire surface properties. Clearly, thermal radiation will serve to lower the wire temperature, so this additional step is necessary to the model. To obtain an upper bound on the radiation, we will assume that the wire surface is a black body and that the surroundings, at the air temperature, are large enough to be effectively black. To keep the solution simple, we will neglect the convection from the wire surface. The radiation energy balance is

$$\sigma \left(T_{\text{wire}}^4 - T_{\text{air}}^4\right) = \frac{Q}{A_{\text{wire}}} \tag{1.7}$$

where σ is the Stefan-Boltzmann constant. In this case, the wire temperature is 655 K or 382°C. Although the wire is still hot, it is at a far more reasonable level than the result estimated using convection alone. Clearly, radiation heat transfer is much more important than convection for this case, justifying our simplifying assumption to neglect convection in Eq. (1.7).

The Second Law of Thermodynamics

The Second Law of Thermodynamics gives a useful upper limit for power generation, propulsion, and air conditioning processes. If the maximum and minimum temperatures for a process can be established, a good estimate for the maximum efficiency follows from the second law.

Example 1.4 *Microturbine power cycle efficiency* Some inventors developed a high-efficiency microturbine expander that used refrigerant vapor as the working fluid. This turbine was the key component to a new compact power supply. However, when combined in a Rankine cycle process, the overall thermodynamic efficiency of the prototypes was very low. The inventors were concerned that the heat exchangers were not properly designed. What they failed to consider is that the refrigerant working fluid has a practical upper temperature limit of about 70°C, as a result of the high saturation pressure associated with this temperature. A power generating cycle using the ambient as the cold reservoir can never exceed the Carnot efficiency associated with these temperature limits.

Taking 20°C as the lower temperature, the maximum efficiency is about 15%. No amount of improvement to the heat exchangers or turbine will allow the device to exceed this limit.

Analogies

Often, physical phenomena in heat transfer or fluid mechanics have close analogies to processes in other fields. The phenomena may be very similar, for example, as the diffusion of heat and mass are similar, or the form of the governing equations may be nearly identical. In such cases, known solutions or experimental results from the analogous field may be used in the problem at hand. Simplified approximations for view factors in the field of lighting have been carried over to thermal radiation. Materials with a high electrical conductivity can be assumed to have a good thermal conductivity because both processes are aided by the mobility of electrons. Sometimes analogies between fields can be accomplished by judicious assumptions or redefinition of the governing parameters. For low mass flow rates, the mass transfer from a solid surface to a flowing fluid can be closely approximated using correlations for convective heat transfer through a suitable analogy between dimensionless parameters. Considerable progress in engineering has occurred through such cross-fertilization.

Example 1.5 *Glass fiber cooling example revisited* Continuing the glass fiber cooling example from before, a more detailed estimate of the glass surface area required for the prescribed glass temperature change can be obtained. The process is analogous to a cross-flow heat exchanger with one stream, the glass, unmixed, and the second stream, the air, approximately well mixed.

Because the air temperature change should remain much smaller than the glass temperature change, the glass has the minimum heat capacity rate, defined as the product of mass flow rate and specific heat. Using results from heat exchanger analysis, the effectiveness, ε, can be estimated from its definition:

$$\varepsilon = \frac{Q}{Q_{\text{ideal}}} \tag{1.8}$$

$$\varepsilon = \frac{(\dot{m}c_p)_{\text{glass}} \Delta T_{\text{glass}}}{(\dot{m}c_p)_{\text{glass}} (T_{g\,\text{in}} - T_{\text{air in}})} = \frac{500}{1000 - 20} \tag{1.9}$$

and $\varepsilon \sim 0.5$ with $(\dot{m}c_p)_{\text{air}} > (\dot{m}c_p)_{\text{glass}}$

From the solution for a cross-flow heat exchanger, the nondimensional surface area, in terms of the number of transfer units (NTU) is

$$\text{NTU} = \frac{hA_{\text{glass}}}{(\dot{m}c_p)_{\text{glass}}} \approx 0.7 \tag{1.10}$$

The required glass surface area A_{glass} can then be found if the glass-to-air heat transfer coefficient is known.

Governing Equations

The governing equations for certain situations are fairly straightforward to write, although their solution may still be difficult. The governing equations can be used to make order of magnitude comparisons between different terms and to identify the controlling phenomena.

By nondimensionalizing the equations, governing dimensionless parameters can also be identified. These can be used to generalize experimental results or guide numerical or experimental investigations by indicating which combination of parameters should be varied. Also, in nondimensional form, analogies to other physical phenomena may become more obvious.

Simplify the Real Case to Known Solutions

Solutions for wide range of thermofluid problems are available in the literature. Suitable simplifications may reduce the model for a complicated device to one for which a known solution exists. Some ingenuity may be needed to see the one-to-one equivalence. In some instances, contributing phenomena can be estimated to exert a second-order effect, so that they can be ignored. The geometry or time-varying conditions can be simplified to the known solution. For example, some two-dimensional conduction problems can be reduced to the case of a one-dimensional fin. Convection with combined forced and natural convection can be reduced to just one form of convection when it can be shown that, say, forced flow dominates over buoyancy driven flow. When such simplifications are achieved, it is important to identify whether they represent an upper or lower bound to the actual situation.

Combined Experiments/Analysis

Simple experiments can be used to identify governing parameters and simplifications. For example, running a given process in a vacuum and noting changes in operating temperatures or heat flux can determine the importance of convection in the real case. Because thermal radiation has a nonlinear temperature dependence, experiments at several different temperature levels can distinguish its influence relative to conduction and convection, both of which vary linearly with temperature difference (assuming the thermophysical properties such as thermal conductivity do not vary significantly with temperature). If radiation can be eliminated by experiment, the resulting heat transfer relations will be linear in temperature, and a simpler solution may be sought.

Graphical Approximations

Before the proliferation of numerical computation, engineering solutions were often obtained by ingenious graphical techniques. Although the more involved of these techniques sometimes required lengthy work and are not at all appropriate today, simple graphical approximations may still provide substantial insight into physical processes. An example is the sketch of isotherms and adiabatic lines in multidimensional conduction problems. These sketches help visualize the physical process and can lead to a simple numerical estimate. An example also occurs in radiation heat transfer, where the complicated geometric relationship between objects can sometimes be reduced to a sketch of strings attached to the ends of the neighboring objects followed by an estimate of their length.

Integral Methods

As an alternative to directly solving differential equations, approximate integral solutions based on energy conservation can provide rather accurate, quick solutions in some cases. The integral solutions require more physical insight into the problem, such as the use of a realistic functional form of the temperature distribution. When flow is involved, as for convective heat transfer over a surface, the integral solution also requires a realistic form for the fluid velocity profile adjacent to the surface.

Numerical Solutions

Numerous computer programs are available to deal with problems involving heat transfer, fluid flow, complicated geometry, and boundary conditions that may include solid, heat conducting structures. Sophisticated packages of this type usually involve a considerable learning curve. Thus, for the uninitiated, use of the approximate methods mentioned previously may be a preferable first step. If the user has familiarity with a straightforward numerical scheme, a good start is to use this on a simplified representation of the case at hand. When a more detailed numerical program is used, it is essential that an approximate or exact solution to a simplified form be developed as a check for proper formulation in the more exact numerical program. This will also help identify the bounds of accurate application of the program.

To obtain an accurate numerical solution for a given problem may require considerable effort to properly represent all of the relevant physical conditions within the numerical program. By suitable simplifications, approximate and bounding solutions can often be obtained much more rapidly. Usually, this requires consideration of order of magnitude estimates and proper bounding calculations before the numerical solutions are undertaken. The proper formulation of boundary conditions and the proper location of boundaries require considerable care.

1.8 Other Factors to Be Considered

Redesign

Although redesign is not properly an approximation technique, it must be kept in mind that the existing device or proposed design under investigation may be inappropriate for the proposed concept. Rather than expending considerable effort to analyze or predict the exact performance of this design, it may be much more valuable to use the approximate estimates and order of magnitude results to identify the limitations of the existing device or the proposed design. The appropriate next step may be the consideration of a modified or totally different design or component.

Property Values

In the same spirit, a redirection of the effort to solve a problem may be in order. For example, rather than spending considerable effort to develop a highly accurate computational or analytical technique for a phenomenon strongly dependent on some physical property that is poorly known, the effort should be refocused to secure better values for the properties. There are numerous cases in which a program embarking on the development of a reliable, accurate prediction of a process had to be redirected toward the measurement of essential property values that were poorly known at the outset. Researchers must be adept at using experimental means as well as computational ones. The determination of the time-varying insulation properties of closed-cell foam mentioned in Section 1.5 is a case in point.

1.9 Summary

In this chapter, the engineering approach for defining a problem was discussed along with various methods of developing simple models. In every case, it is essential to begin by understanding the goals of a modeling effort – what question is to be answered. In the initial design stages, it is often necessary to estimate system performance only well enough to understand how the design compares to other possible designs or whether its performance is even in the right ballpark relative to the desired outcome. As such, initial models often need not go to high precision and should instead be focused on capturing the essential physics with limited numerical precision and in a relatively short span of time. Several examples of this approach were given in the present chapter, and a number of basic methods were described. These included making order of magnitude estimates, establishing upper and lower bounds on the real system, using the governing equations to identify controlling phenomena and drawing analogies to know cases. These methods as well as a number of additional ones will be developed in more detail in subsequent chapters.

BIBLIOGRAPHY

S. J. Klein. *Similitude and Approximation Theory*. New York: McGraw-Hill, 1965.
J. H. Lienhard IV and J. H. Lienhard V. *A Heat Transfer Textbook*, 4th ed. Mineola, NY: Dover Publications, 2011. Free ebook (http://ahtt.mit.edu).
A. F. Mills. *Heat Transfer*, 2nd ed. Upper Saddle River, NJ: Prentice Hall, 1999.

W. M. Rohsenow and H. Y. Choi. *Heat, Mass, and Momentum Transfer.* Englewood
    Cliffs, NJ: Prentice-Hall, 1961.
H. Schlichting. *Boundary-Layer Theory*, 6th ed. New York: McGraw-Hill, 1968.

PROBLEMS

1.1 In a manufacturing process for ceramic fibers, the fiber surface must be protected
by a thin coating. The coating is applied to a large bundle of fibers in a dilute
water solution. The water-saturated bundle must then be dried. One proposal to
dry the bundle of fibers involves pulling them over a very hot steel plate, as shown
in the diagram. Each fiber in the bundle has a diameter of approximately 0.5×10^{-3} inches (0.012 mm). The bundle of fibers moves at a velocity of 20 ft/s (6 m/s)
vertically downward adjacent to the plate. The fibers are close spaced within the
bundle. The cross section of the bundle is 0.1 inch wide and 0.015 inches thick
(2.5×0.38 mm). The 0.1-inch side is adjacent to the hot steel plate. The total
plate height is 1 ft (0.3 m), and the plate temperature is 1300°F. The wet fibers
approaching the plate are at a temperature of 100°F.

   a. If the heated plate successfully dries the fibers, how important is radiation as
   opposed to conduction and convection for the plate-to-fiber heat transfer?

   b. Estimate the order of magnitude of the fluid layer between the plate and the
   fibers.

   c. How could you test whether the layer between the plate and the fibers is liquid
   or vapor?

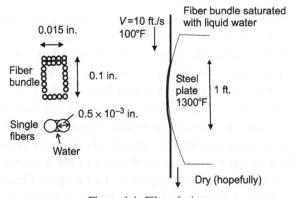

Figure 1.4. Fiber drying.

1.2 A company proposes to use a reciprocating steam engine of 1000 HP output
to power a medium size helicopter (a real story). Making upper limit estimates
of cycle performance, and temperatures, estimate the minimum condenser heat
transfer rate for a closed steam cycle, the surface area of an air cooled condenser,
and the frontal area for air flow through the condenser. Will it fly?

1.3 One means proposed to conserve energy for space heating is the use of night
setback: the interior temperature of a building is reduced during the evening.
The heat transferred from the building to the outside air is directly proportional to the temperature difference between the building interior and the

outdoor ambient temperature. Between the hours of 8 AM and 10 PM the interior temperature in the winter should remain at 20°C. At night for the remaining 10 hours, it cannot fall below 12°C when the ambient temperature is constant at 0°C. Some people have questioned this strategy, claiming that the total energy saved by the night setback is reduced because of the additional energy needed to raise the interior temperature in the morning. Can you estimate the impact of this on the savings resulting from the night setback for a typical single-family home in the Boston area?

1.4 Heard on National Public Radio:

Most Americans use electricity, gas, or oil to heat and cool their homes. But the small city of Brainerd, Minnesota is turning to something a bit less conventional: the sewer. As it turns out, a sewer – the place where a city's hot showers, dishwashing water, and organic matter end up – is a pretty warm place. That heat can generate energy – meaning a city's sewer system can hold tremendous potential for heating and cooling. It's just that unexpected energy source that Brainerd hopes to exploit. Scott Sjolund, technology supervisor for Brainerd Public Utilities, is standing on the corner of 6th Avenue and College Drive in Brainerd, as sewage rushes unseen through underground pipes. "Everybody heats water up ... and all that gets drained down the sewer, and that's potential energy that could be extracted. That's part of the equation," (Published July 12, 2012 by Conrad Wilson). For a community in Minnesota, could the heat from its sewage provide a significant portion of the population's home heating needs?

Steady-State Conduction Heat Transfer

2.1 Introduction

From a mechanistic point of view, there are two basic forms of heat transfer: conduction and radiation. Other heat transfer processes such as convection, boiling, and condensation involve conduction aided by fluid motion (and sometimes radiation). Our discussion of heat transfer modeling starts with the case of steady-state conduction without flow.

Although steady conduction is normally regarded as a straightforward case, in numerous instances modeling is required and can be helpful. In particular, the electrical analogy between one-dimensional heat conduction and electric current flow allows an easier visualization and can lead to more straightforward solutions. Often this involves simplifying the geometry of a problem. Some two-dimensional cases can be closely modeled using the fin solution. Several means to model two- and three-dimensional conduction are described in this chapter, ranging from circuit models, to graphical approximations, to the use of known solutions for specific geometries. Beyond the geometry, convective and radiative boundary conditions have a substantial influence on overall heat transfer while bringing in another level of uncertainty. These ambiguities may be heightened by less than complete knowledge of material composition and the resulting thermophysical properties.

Molecular View

From a molecular point of view, conduction occurs when a temperature difference is present across a solid, liquid, or gas. By molecular motion, energy is transferred from the warm, more energetic molecules or atoms to the cooler, less energetic molecules. In solids, the energy is transferred by lattice vibrations called phonons and free electron motion. In gases, the energy is transferred by the motion of relatively widely spaced molecules. The energetic molecules collide with less energetic ones and transfer energy to them. In liquids, the molecular spacing is much closer but again the energy transfer takes place by virtue of molecular motion. Note that heat transfer is energy transfer by virtue of a temperature difference. The flow of a warm fluid out of a volume of space represents an energy transfer but not necessarily a heat transfer.

The molecular view of conduction heat transfer is useful to gain a physical understanding of the process. As a good rule of thumb, a material that is a good electrical conductor is also a good conductor of heat. Similarly, gases with a low molecular weight have a higher conductivity than gases with complex molecules. The molecular approach can be used to predict microscopic thermophysical processes and properties. However, the molecular approach is usually not a fruitful means to model a large-scale complex macro process when the physical dimensions are much larger than the mean free path.

Macroscopic Approach

For this work, a macroscopic phenomenological approach is followed. We begin with a brief review of basic principles. The rate of heat transfer is found to be proportional to the difference or gradient of a driving potential, the temperature. In vector notation the phenomenological expression for the rate of heat transfer $\vec{Q}$, in W or BTU/hr, is

$$\vec{Q} \propto - A\mathrm{grad}\,T \qquad (2.1)$$

where A is the area normal to the gradient of the temperature, T. Written as an equality, it is Fourier's Law,

$$\vec{Q} = -kA\mathrm{grad}\,T \qquad (2.2)$$

where k is the thermal conductivity, a thermophysical property of a homogeneous single component material.

2.2 Property Values

For pure single-component materials, such as pure metals, the thermal conductivity is well known, at least at moderate pressures and temperatures. Many engineering materials are complex combinations or alloys of several elements. For these substances, the conductivity may vary substantially with relatively small changes in composition. Other engineering materials offer similar challenges in property characterization, such as composite materials, polymers, and porous materials.

A common method to measure k is to contain a thin sample of material between two parallel flat plates that are kept at two different constant temperatures, T_1 and T_2. The two temperatures are measured along with the rate of heat transfer. The process is allowed to reach steady state, so that the temperature at every point in the body does not vary over time. If properly set up (more about this later), the temperature varies only in the direction normal to the plate surfaces and the heat transfer is one-dimensional.

In this case, Fourier's Law, written for any surface through the body parallel to the plate surface, becomes

$$Q = -kA \frac{dT}{dx} \tag{2.3}$$

In steady state, without any sources of energy in the body, such as a chemical reaction, an energy balance on a plane layer of thickness Δx indicates that Q must be the same on both surfaces of the layer. The solution to Eq. (2.3) when the conductivity does not vary through the sample thickness, w, is

$$-kA \frac{\Delta T}{\Delta x} = +kA \frac{T_2 - T_1}{w} = Q = \text{constant} \tag{2.4}$$

Thus, if T_1 and T_2 are measured along with Q, k can be found. For gases and liquids convective motion must be suppressed. For all materials, the boundary conditions on the sides as well as the top and bottom must be carefully controlled.

The same experimental procedure can be used to measure the conductivity of a heterogeneous material such as foam or fibrous insulation. Wood, formed of a cellular structure, is one such material. In this case the heat transfer takes place by a combination of conduction and possibly radiation through the solid and conduction and radiation through the gas filling the voids within the solid. Notwithstanding these complexities, the experiment, along with Eq. (2.3), gives a value for k that is referred to as the effective or overall conductivity.

For heterogeneous materials, the morphology of the material as well as the material composition influences the effective conductivity. Some closed-cell insulation foams are made with high molecular weight gases inside each cell. The effective conductivity of these foams can change by 30% or more over the life of the material because of the diffusion of gases into and out of the foam's interior. For some materials, the effective conductivity may include substantial heat transfer by infrared radiation. For these, the measured effective conductivity may also vary with sample thickness and emissivity of the bounding surfaces.

Figure 2.1 shows the range of k values for insulation, wood, plastic, concrete, steel, and copper (both pure and alloyed). Note that the scale is logarithmic. For wood, which is anisotropic, the orientation of the wood grain relative to the temperature gradient and the percent of absorbed moisture can each have as much as a factor of 2 influence on the conductivity.

Not only can there be strong variations in conductivity with material composition, but there are also uncertainties in the conductivity measurement. These can include convective motion in gases and liquids as well as in gas-filled voids of heterogeneous materials. Radiation heat transfer in gas conductivity cells and in semi-transparent solid materials such as glass or plastic can cause erroneous conductivity values. One must carefully assess the data source before reported conductivities can be used with confidence.

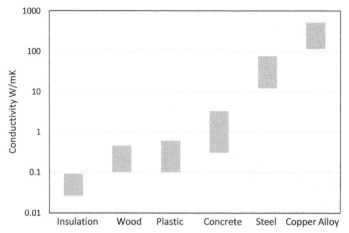

Figure 2.1. Range of thermal conductivity. The shaded portion shows the typical range of values for that material.

2.3 Electrical Analogy

The electrical analogy to conduction is a useful tool to visualize the heat transfer. We begin with elementary concepts and definitions. The expression for one-dimensional steady heat conduction is similar to Ohm's Law for DC electric current flow,

$$I = \left(\frac{\Delta V}{R_e}\right) \tag{2.5}$$

For a one-dimensional body of thickness X steady conduction can be written in a similar form,

$$Q = kA\left(\frac{\Delta T}{X}\right) = \left(\frac{\Delta T}{X/kA}\right) = \left(\frac{\Delta T}{R_T}\right) \tag{2.6}$$

For the electrical analogy to conduction, Q is analogous to current I and the driving potential for conduction, ΔT, is analogous to the voltage difference, ΔV. Finally, X/kA is analogous to the electrical resistance, R_e, and is defined as the thermal resistance R_T.

It is helpful to represent this graphically by an equivalent circuit shown in Figure 2.2.

To compare products used for insulation, the R factor, defined as $R_T A$ or X/k, is often used. Thus the R factor is independent of the surface area and is a better measure of the material's intrinsic performance. Note, the R factor, R_f, is given in a variety of different units.

Figure 2.2. Electrical analogy.

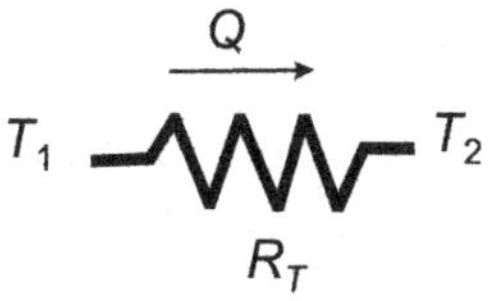

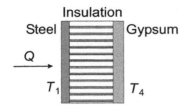

Figure 2.3. Resistances in series.

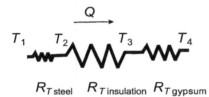

Example 2.1 The steady-state performance of a system consisting of many materials in layers such as the wall of a steel-clad building can be approximately represented by a DC thermal circuit. In steady state with the conduction through each element one-dimensional the heat transfer rate through each element in Figure 2.3 is the same. The overall heat transfer rate is the ratio

$$Q = \frac{T_1 - T_4}{R_{\text{T steel}} + R_{\text{T insulation}} + R_{\text{T gypsum}}} \tag{2.7}$$

$$Q = \frac{T_1 - T_4}{\left(X/Ak\right)_{\text{steel}} + \left(X/Ak\right)_{\text{insulation}} + \left(X/Ak\right)_{\text{gypsum}}}$$

$$= \frac{A\left(T_1 - T_4\right)}{R_{f\text{steel}} + R_{f\text{insulation}} + R_{f\text{gypsum}}} \tag{2.8}$$

In this example, the size of the resistors shown in Figure 2.3 represents the relative magnitude of the corresponding thermal resistance. For elements in series, the element with the largest resistance has the strongest influence on the overall heat transfer rate. Note, when using this approach one must confirm that the heat transfer is indeed close to one-dimensional. This question will be taken up later. For the remainder of the book, we will represent thermal resistance as simply R and omit the subscript T, but we will retain R_f as the R factor.

2.4 Approximate Estimates of Magnitude

In some instances, simple order of magnitude estimates are sufficient to determine the value of novel designs. Several examples follow that illustrate the use of this approach coupled with limit analysis.

Example 2.2 *Heat transfer through windows* One-dimensional heat flow should occur through a single-pane window (except near its edges). At first glance this would suggest that changing the window material from glass to plastic would reduce the heat flow by roughly a factor of 5, the ratio of glass to plastic thermal

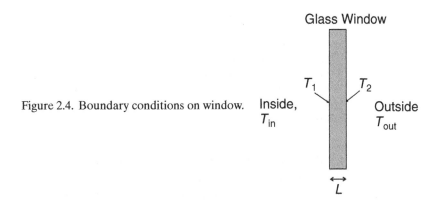

Figure 2.4. Boundary conditions on window.

conductivity. However, to deal with this example properly, the boundary conditions on the window must also be examined. The difference between the ambient air temperature outside the window and the interior air temperature, not the window surface temperatures, is the driving potential that remains the same when the window material is changed.

At the interior glass surface, the thermal resistance from the interior to the surface of the window must be accounted for. This involves convection and radiation heat transfer.

Convection and radiation are dealt with in detail in later chapters. The approximate order of magnitude of these resistances is introduced here.

Convection Heat Transfer

Heat transfer from a surface to a surrounding fluid is due to conduction augmented by mass and energy transfer set up by fluid motion. The process is termed convection heat transfer. There is a thin boundary layer in the air adjacent to the surface within which the air temperature variation is confined. The rate of heat transfer is proportional to the surface area and the temperature difference between the surface and the uniform air temperature outside the boundary layer. The coefficient of proportionality, called the heat transfer coefficient, h, is defined by

$$Q = hA(T_{\text{surface}} - T_{\text{fluid}}) \tag{2.9}$$

where h has the units of $W/m^2\,K$ (or $BTU/hr\ ft^{2\circ}F$). Of course, Eq. (2.9) is not of much use until some way to predict or estimate the value of h is established.

Generally, the heat transfer coefficient, h, is a function of the fluid properties, the fluid velocity, the surface geometry, and sometimes the temperature level. A more detailed discussion of convection is given later. For now it is sufficient to observe that h increases as the air velocity increases and that it is larger for fluids of higher thermal conductivity.

Convection has two general forms. When the air motion is set up by buoyancy effects due to the temperature difference between the surface and the fluid, such as for the air flow over a hot "radiator," the flow is called natural or free convection.

Table 2.1. *Convection heat transfer coefficients* (W/m^2K)

Gases, natural convection	 5–30
Gases, forced convection	 10–250
Liquids, natural convection	 15–1000
Liquids forced convection	 150–5000
Boiling liquids	 1000–250,000
Condensation	 2500–25,000

When the flow is due to an external source, such as the wind, a fan, or motion of the surface itself, the flow is forced convection.

Rohsenow and Choi [2.1] have presented a table that gives useful estimates of the order of magnitude of h for convection heat transfer as well as for boiling and condensation heat transfer. It is reproduced as Table 2.1. Note that for objects with small dimensions, millimeter scale or less, h values can be much larger than those given in the table.

Now, returning to the concept of thermal resistance, from Eq. (2.5) the equivalent thermal resistance, R_T, for convective heat transfer is $(1/hA)$. An equivalent R factor for convection is then $1/h$.

Radiation Heat Transfer

Radiation is energy transfer between bodies at different temperatures due to electromagnetic radiation. It differs from conduction in that it does not require an intervening molecular medium for the energy transfer.

Black bodies absorb all of the radiation that is incident on their surface. When two parallel black body planes are each at a uniform temperature, T_1 and T_2 respectively, the medium between them is transparent, and the plate spacing is small compared to the plate length (details of this later), then the net radiation between the two plates is given as

$$Q = \sigma A \left(T_1^4 - T_2^4\right) \tag{2.10}$$

where σ is the Stefan–Boltzmann constant, equal to 5.7×10^{-8} W/m^2 K^4 (1.7×10^{-9} BTU/hr ft$^{2\circ}$R^4), and the temperatures T_1 and T_2 are in the absolute scale (kelvin or degrees Rankine).

When the difference between T_1 and T_2 is not large, Eq. (2.10) can be approximated as

$$Q = 4\sigma A (T_m)^3 (T_1 - T_2) \tag{2.11}$$

where T_m is the mean temperature in the absolute scale (the limits of this approximation are dealt with in Chapter 6). Equation 2.11 can be written in terms similar to the equation for convection heat transfer, Eq. (2.9), by use of a fictitious, but useful, radiative heat transfer coefficient, h_r:

$$Q = h_r A (T_1 - T_2) \tag{2.12}$$

Table 2.2. *Values of radiative heat transfer coefficient h_r for black bodies*

Mean temperature T_m (K)	h_r (W/m²K)
273	4.7
300	6.2
600	50
900	170
1200	400

Table 2.2 gives typical values of h_r, which can be compared to the values of convective heat transfer coefficients of Table 2.1.

For a surface in contact with air, which can be assumed to be transparent, radiation and convection to the surroundings act in parallel. In this range of temperature (for which radiation is mostly in the mid and long infrared wavelengths), glass is very nearly a black body. The electrical analogy for the window shown in Figure 2.4 is given in Figure 2.5 where the radiative surface temperatures of the room and the outside are assumed to be the same as the corresponding interior and exterior air temperatures, respectively.[1]

The effect of changing the window from glass to plastic must be evaluated in terms of the entire series of resistance between the interior and exterior air.

For radiation and convection acting in parallel,

$$Q = (h_r + h_c)A_{in}(T_{in} - T_1) \qquad (2.13)$$

These effects can be combined into a single thermal resistance as follows

$$R = \frac{1}{(h_r + h_c)_{in}A} \qquad (2.14)$$

These combined resistances lie in series with the conduction resistance of the window material. The heat transfer rate is the ratio of the overall temperature difference to the sum of the series resistances:

$$Q = \frac{T_{in} - T_{out}}{\dfrac{1}{(h_r + h_c)_{in}A} + \left(\dfrac{X}{KA}\right)_{glass} + \left(\dfrac{1}{h_r + h_c}\right)_{out}A} \qquad (2.15)$$

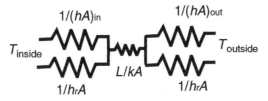

Figure 2.5. Electrical analogy for heat transfer through window.

[1] Note in the infrared, glass is opaque and approaches black body characteristics.

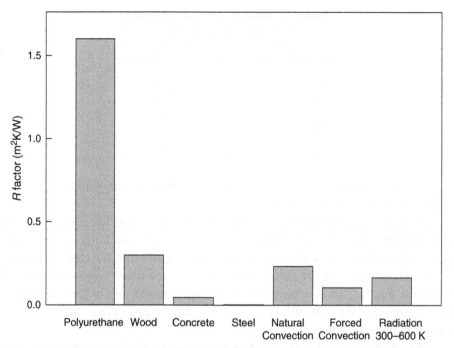

Figure 2.6. *R* factors for 2-cm thick solids, convection, and black body radiation. Shaded regions show the range of values.

A simple order of magnitude estimate can be done with A omitted as it is a common factor to each term. In other words, we can compare with the R factors

$$\frac{1}{(h_r + h_c)_{\text{in}}} \sim \frac{1}{(5 + 5)} \sim \frac{1}{(10)} \left(\frac{W}{m^2\,K}\right)^{-1}$$

$$\frac{1}{(h_r + h_c)_{\text{out}}} \sim \frac{1}{(5 + 10)} \sim \frac{1}{(15)} \left(\frac{W}{m^2\,K}\right)^{-1} \qquad (2.16)$$

$$\left(\frac{X}{K}\right)_{\text{glass}} \sim \frac{5 \times 10^{-3}}{1} \sim 5 \times 10^{-3} \left(\frac{W}{m^2\,K}\right)$$

The resistance of the glass is at least an order of magnitude less than the convective and radiative resistance on the inside and outside surfaces. If plastic is substituted for glass, the R factor of the solid would increase by about a factor of 5 but the total heat transfer through the window will be decreased by only about 10%. There is no need to do a more detailed analysis to pin down more accurate estimates of convection or radiation: these estimates show that the improvement in window performance (at least in terms of heat loss) is marginal at best.

To generalize the above comparison, Figure 2.6 illustrates the approximate R factors (X/k) of 2 cm thick solids and polyurethane foam as well as the R factors for convection and thermal radiation for black bodies with a mean temperature between 300 K and 600 K.

Example 2.3 *Dropwise condensation* To enhance the performance of condensers used in power plants or refrigeration equipment, designers would like to use a special form of condensation known as dropwise condensation. What level of improvement can be expected? Condensation on a typical metal surface occurs by film condensation, where the condensate wets the metal surface and spreads out to form a continuous liquid layer. The heat transfer coefficient for film condensation is on the order of 10,000 W/m² K. If the surface is treated so that the liquid does not wet it, it forms discrete drops rather than a continuous film; the condensation heat transfer coefficient is enhanced and can reach levels of 200,000 W/m² K, resulting in a much lower thermal resistance.

The problem is maintaining the nonwetting characteristic of the surface. Condensing vapor continuously moves to the surface and impurities in the fluid system are brought to the surface, where they contribute to liquid wetting. One permanent solution is to coat the metal with a nonwetting polymer (such as Teflon®). If such a polymer coating is 20 μm thick, the resistance, using a typical conductivity for polymers of 0.2 W/m K, becomes

$$\frac{X}{k} \sim \frac{20 \times 10^{-6}}{0.2} \sim 10^{-4} \left(\frac{W}{m^2 K}\right)^{-1} \tag{2.17}$$

whereas the resistance of a typical film condensation is

$$\frac{1}{h_{\text{film}}} \sim \frac{1}{10^4} \sim 10^{-4} \left(\frac{W}{m^2 K}\right)^{-1} \tag{2.18}$$

Therefore, the resistance of the coating to maintain dropwise condensation is roughly the same as the resistance for film condensation, thereby eliminating all of the thermal performance advantage of dropwise condensation. For dropwise condensation to be advantageous, much thinner polymer coatings or higher conductivity surface coatings are required. Currently, researchers are developing nonwetting surfaces with nanoscale structures to meet these challenges.

Example 2.4 *Gas turbine blade cooling* To increase the efficiency and the net power-to-weight ratio in a gas turbine the maximum cycle temperature must be increased. The upper temperature is limited by the conditions on the rotating turbine blades at the exhaust of the combustor. The maximum material temperature must be below the melting point of the metal alloy. The rotating turbine blades are hollow, and cool air from the compressor passes through the interior of the blade. Since the blades rotate rapidly, the exterior gas temperature can be taken as the spatial average temperature of the combustor exhaust. The heat transfer process from the combustion gas to the cooling interior air can be considered steady state.

A simple resistance network for an element of the blade thickness has the exterior, metal, and interior resistances in series. Heat is transferred from the hot gas to the exterior surface of the turbine blade by convection and radiation. We can estimate the thermal resistance using Tables 2.1 and 2.2:

$$R_{\text{exterior}} \approx \left[\frac{1}{h_{\text{exterior}} + h_r}\right]\frac{1}{A} \approx \left[\frac{1}{100 + 170}\right]\frac{1}{A} \approx \frac{4 \times 10^{-3}}{A}[\text{W/m}^2\text{K}]^{-1} \quad (2.19)$$

For the turbine blade with a thickness of 5 mm and a conductivity of about 40 W/m K the associated thermal resistance is

$$R_{\text{metal}} \approx \frac{X}{kA} \approx \frac{5 \times 10^{-3}}{40A} \approx \frac{10^{-4}}{A}[\text{W/m}^2\text{K}]^{-1} \quad (2.20)$$

Heat is transferred from the interior blade surface to the cooling air by convection. The associated R value is

$$R_{\text{interior}} = \frac{1}{hA_{\text{Interior}}} \approx \frac{1}{100A_{\text{Interior}}} \approx \frac{10^{-2}}{A_{\text{Interior}}}[\text{W/m}^2\text{K}]^{-1} \quad (2.21)$$

In this case, the resistance of the metal thickness is negligible compared to the exterior and interior resistances. The key to maintaining the metal at a safe temperature is to decrease the internal heat transfer resistance. In practice, this is done by adding ribs and extended surfaces to the interior surface to increase the heat transfer coefficient and adding small fins to increase the internal surface area.

Example 2.5 *Glass fiber insulation* Glass fiber insulation, sometimes referred to as glass wool or fiberglass, consists of a matrix of fine glass fibers that resembles cotton candy. The size and number of glass fibers is designed to set up enough flow resistance to suppress convection in the air between the fibers. The effective conductivity of the insulation has been measured to have a value of roughly 0.05 W/m K at room temperature. At the same temperature air has a conductivity of 0.026 W/m K. To explain this difference, the role of conduction through the glass fibers must be examined. The glass fiber insulation has a density of about 16 kg/m^3.

An upper limit to the contribution of conduction through the glass can be obtained by assuming all of the glass is aligned parallel to the direction of heat flow. The solid glass has a density of 2500 kg/m^3 and a conductivity of approximately 1 W/m K. The volume fraction of glass is the ratio 16/2500 or 0.006. This also represents the fraction of the cross-sectional area occupied by the parallel glass fibers. The total heat transfer by conduction through the air and glass may be written as

$$Q = (k_{\text{glass}}A_{\text{glass}} + k_{\text{air}}A_{\text{air}})(\Delta T) = k_{\text{effective}}A_{\text{total}}\Delta T \quad (2.22)$$

Dividing by the product of A_{total} and ΔT yields one estimate for the effective conductivity of the insulation,

$$k_{\text{effective}} = 0.994 \times 0.026 + 0.006 \times 1 \approx 0.032 \text{ W/mK} \quad (2.23)$$

This value is well below the experimentally measured value of the insulation, even though the maximum possible glass conduction was assumed. In the real

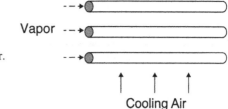

Figure 2.7. Air-cooled condenser.

Cooling Air

case, most of the glass is aligned roughly perpendicular to the direction of heat transfer. This upper limit estimate indicates two things: the conduction through the glass is not substantial and there must be another heat transfer mechanism in play. Assuming that convection within the air is suppressed (which is generally the case) then the only other mechanism must be thermal radiation. Moreover, comparing the effective conductivity with the air conductivity, heat transfer by thermal radiation must be roughly equivalent to conduction through the air in the glass fiber insulation. Experiments have confirmed this. When the glass fibers are made thinner, insulation with the same mass density has more glass surface area; it becomes more opaque and the effective conductivity is reduced.

The minimum contribution the glass can make occurs when all of the glass fibers are perpendicular to the direction of heat flow. Assume the glass fibers form a solid layer of thickness X_{glass} over the insulation cross-section, so that the glass and the air within the insulation are in series. Then, the total resistance can be determined as the sum of resistance of the glass and the air,

$$\frac{X_{total}}{k_{effective-min}} = \frac{X_{glass}}{k_{glass}} + \frac{X_{air}}{k_{air}} \approx \frac{0.006}{1} + \frac{0.994}{0.026} \approx \frac{0.994}{0.026} \qquad (2.24)$$

and $k_{effective-min}$ neglecting radiation is almost identical to k_{air}.

Example 2.6 *Air-cooled condenser* Simple upper and lower limit approximations can also be used to estimate the order of magnitude of the size of a heat exchanger. Consider an air-cooled condenser made up of a series of parallel cylindrical tubes. In this design, vapor is condensed within the tubes as cooling air flows over the tube surface as shown in Figure 2.7.

As in previous examples, it can be shown that the resistance of the tube wall is small compared to the convection resistance on the outside of the tube. The order of magnitude of condensation heat transfer coefficients is much larger than single-phase convective flow of air (see Table 2.1). Thus, the major resistance is due to convection on the outside of the tube. In the limiting case of a high air flow rate, the temperature change of the air through the exchanger is small. The pressure drop of condensate is usually small, allowing the saturation temperature within the tube to be approximated as constant. In this limiting case, both the inside and outside temperatures are constant for the entire heat exchanger. The heat transfer rate can be expressed as

$$Q \approx h_{outside} A_{surface\ MIN} (T_{saturation} - T_{air\ entering}) \qquad (2.25a)$$

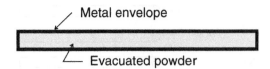

Figure 2.8. Evacuated panel.

where Q is found from the product of the vapor mass flow rate and the heat of vaporization. $A_{\text{surface MIN}}$ represents the minimum surface area required for the heat exchanger.

The upper limit for the required surface area (while T_{sat} remains constant) can be found by using the exit air temperature in Eq. (2.25a),

$$T_{\text{air exit}} = T_{\text{air entering}} + \frac{Q}{(\dot{m}c_p)_{\text{air}}} \tag{2.25b}$$

In that limit Eq. (2.25b) relates Q to the maximum, upper limit, required surface area,

$$Q \approx \frac{h_{\text{outside}}A_{\text{surface MAX}}(T_{\text{saturation}} - T_{\text{air entering}})}{1 + \dfrac{h_{\text{outside}}A_{\text{surface MAX}}}{\dot{m}c_p}} \tag{2.25c}$$

Example 2.7 *Conduction in parallel panel, vacuum insulation* Evacuated panels are a promising way to improve insulation performance for shipping containers, appliances and buildings. In these applications, a flat panel is required. A typical panel contains fibers, powder, or open-cell foam matrix that provides rigidity. The matrix is covered with a thin impermeable envelope and is evacuated. For long life the envelope must be metallic to prevent air diffusion into the matrix. A typical panel is shown in Figure 2.8.

The effective conductivity of evacuated powder has been measured to be 0.5×10^{-2} W/m K. For a square panel $1/2$ m wide and 2 cm thick a stainless-steel envelope $1/10$ mm thick has been suggested. The overall thermal resistance of the panel with the envelope needs to be assessed. The possibility of lateral conduction along the top and bottom face of the envelope, perpendicular to the applied temperature gradient, must be considered. The two limiting cases will be considered for lateral conduction. As a first step, consider the relative magnitude of lateral conduction in the evacuated powder and the metal face. The product of conductivity and cross-sectional area in the lateral direction for the panel can be compared to that in the metal cross-section on one surface

$$\frac{(kA)_{\text{metal}}}{(kA)_{\text{powder}}} \approx \frac{20 \cdot (10^{-4})Z}{(0.5 \cdot 10^{-2})(2 \cdot 10^{-2})Z} = 20 \tag{2.26}$$

where Z is the dimension normal to the plane illustrated. For a first-order consideration, the lateral conduction in the powder can be ignored: only conduction in the steel face might be significant.

An upper limit of panel thermal resistance can be obtained by assuming that lateral conductivity in both the powder *and* the steel envelope is negligible. The

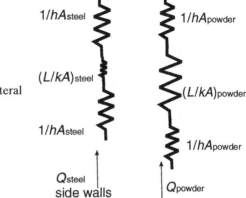

Figure 2.9. Upper limit for R, without lateral conduction.

two faces of the panel will generally be in contact with air for most applications. Both faces will have a convective boundary condition; radiation will be ignored for simplicity. If there isn't any lateral conduction, the heat transfer through the metal sides of the panel and through the powder will act in parallel.

Figure 2.9 shows the electrical resistance circuits for this case. The resistance to conduction through the upper and lower steel envelopes has been neglected. Since all lateral heat transfer is neglected in this limiting case, the heat transferred through the side steel walls is transferred by convection to the air through the same cross-sectional areas as the side walls of the steel. Because this area is small, the resistance of the convective terms is large, limiting the heat transfer through the steel parallel circuit. Using typical values for the convective heat transfer coefficient, it can be shown that the heat transfer through the parallel steel circuit is small compared to that through the much larger cross-sectional area of the evacuated powder when we neglect lateral conduction:

$$L/kA_{\text{powder}} \approx \frac{2 \cdot 10^{-2}}{(0.5 \cdot 10^{-2})(0.5)(0.5)} = 16 \, (\text{W/K})^{-1}$$

$$2/hA_{\text{steel side walls}} \approx \frac{2}{5 \cdot 4 \cdot (0.1 \cdot 10^{-3})} = 1000 \, (\text{W/K})^{-1}$$

(2.27)

A lower limit of panel R_T occurs when the lateral conductance of the top and bottom metal faces of the panel is assumed to be very high. Under these limiting conditions, the temperature of the upper and lower metal faces is uniform in the lateral direction. This occurs when the metal faces are too thick. In this case, the analogous electrical circuit is as shown in Figure 2.10. The side wall conduction resistance is

$$\frac{L}{kA_{\text{steel}}} \approx \frac{2 \cdot 10^{-2}}{20 \cdot 4 \cdot 10^{-4}} = 2.5 (\text{W/K})^{-1}$$

(2.28)

This thermal resistance is much lower than that through the evacuated powder given in Eq. (2.27), substantially compromising the overall panel performance. In this limit, most of the heat bypasses the powder as it flows along the

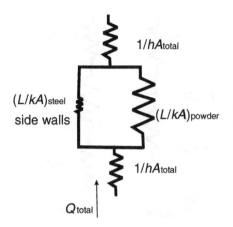

Figure 2.10. Lower limit for R, high lateral conductivity in top and bottom envelope faces.

metal side walls. When it reaches the top of the panel, it flows laterally across the entire top metal face where it is transferred by convection to the air.

Since the upper and lower limits are far apart, a more accurate estimate is required to evaluate the expected success of the panel; the simple estimates are not sufficient. This requires a more careful consideration of two-dimensional heat transfer, addressed in the next section, to determine when conduction along the thickness of the metal faces will compromise the panel performance.

2.5 Two- and Three-Dimensional Steady Conduction

Most real engineering problems involve two- and three-dimensional conduction heat transfer. For example, a body may have a complex shape or the ends of a thin flat plate may have conditions that induce lateral heat conduction from the interior. Numerous numerical codes exist to calculate two- and three-dimensional conduction heat transfer; but complex geometries and boundary conditions can make the computation time consuming. In many instances, it makes more sense to first try approximate techniques to estimate the magnitude of the multidimensional conduction. Later in this chapter, we use such an approximation to further evaluate the vacuum panel discussed in the preceding example.

General Steady-State Equation

For steady state conditions in a body with possible conduction in three orthogonal directions as well as a volumetric heat source $\dot{q}_s$ (see Figure 2.11), the energy equation is

$$\frac{\partial}{\partial x}\left(k\frac{\partial T}{\partial x}\right) + \frac{\partial}{\partial y}\left(k\frac{\partial T}{\partial y}\right) + \frac{\partial}{\partial z}\left(k\frac{\partial T}{\partial z}\right) + \dot{q}_s = 0 \qquad (2.29)$$

Example 2.8 *Conductivity test apparatus* As an example, consider a steady-state method used to measure the conductivity of a material that was mentioned earlier. A thermal conductivity tester is shown in Figure 2.12. To obtain accurate

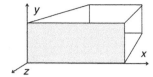

Figure 2.11. Three-dimensional conduction.

results, one-dimensional heat transfer has to be closely established. These testers are designed so that the sample thickness X is much smaller (by a factor of 5 to 10) than the sample width Y or the sample length Z. A sample is placed between two plates maintained at different temperatures, and heat is applied to the upper surface while the lower surface is cooled. The test continues until the sample has reached a steady-state temperature distribution, unchanging with time. The temperature difference between the plates is modest so that the thermal conductivity throughout the sample can be assumed uniform. By measuring the heat transfer rate and the temperature difference across the sample, the conductivity can be determined.

By a rough approximation, the rate of conduction in the x direction is

$$Q_x = kYZ\frac{\partial T}{\partial x} \approx kYZ\frac{\Delta T_x}{X} \tag{2.30}$$

and in the y or z direction,

$$Q_y = kXZ\frac{\partial T}{\partial Y} \approx kXZ\frac{\Delta T_y}{Y} \tag{2.31}$$

If X is small compared to Y and Z, then one would expect $Q_x \gg Q_y$ if the temperature differences, ΔT_x and ΔT_y, in Eqs. (2.30) and (2.31) are the same order of magnitude. How small is "small enough" depends on the accuracy required as well as the boundary conditions on the body being tested. Estimating the temperature gradients in the respective coordinate directions takes some care, to establish whether the approximations given in Eqs. (2.30) and (2.31) will hold. Several different methods will be considered to estimate the lateral heat transfer.

If the sides of the sample are carefully controlled to have a linear temperature gradient from T_1 to T_2, the upper and lower plate temperatures, respectively, one-dimensional heat flow prevails throughout the sample. However, if the sides

Figure 2.12. Thermal conductivity measurement apparatus. z is normal to the x-y plane.

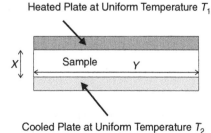

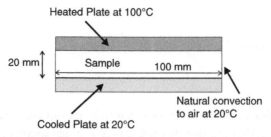

Figure 2.13. Thermal conductivity measurement without insulated sidewalls.

have a distribution other than linear there will be two- or three-dimensional heat flow in the y and z directions.

If the sidewalls are not well insulated or if there is a gap between the side walls and the adjacent insulation, convective heat transfer to air in the gap can occur. It would then be helpful to estimate the upper limit of that heat transfer. When the lateral heat transfer is appreciable compared to the heat transfer in the x direction, substantial errors in the measurements would ensue. Figure 2.13 shows a possible scenario with a 20 mm sample thickness and a sample width of 100 mm. For simplicity, we will assume the z dimension is much larger than these dimensions, so that heat conduction will be two dimensional.

To make the discussion more straightforward, we will take the material to be clamped tightly at the top and bottom to high conductivity plates kept at 20°C and 100°C, respectively. It is reasonable to assume these two boundaries are at uniform temperature. The sides of the material are exposed to air at 20°C. Heat transfer from the exposed sides to the air is by natural convection.

Estimate: Using a Graphical Technique

First, we will use a rough graphical sketch to help establish the resulting temperature field. The sketch, Figure 2.14, will give us a qualitative physical picture of the process. It is not intended to produce accurate quantitative estimates. Near the top and bottom plate isotherms will be nearly horizontal. As heat flows out the sidewalls the isotherms will be distorted. It helps in making such a sketch to remember that adiabatic boundaries (or heat flow lines) will be perpendicular to isotherms since at any point the heat flux vector is proportional to the temperature gradient,

$$\vec{q} = -k \operatorname{grad} T \tag{2.32}$$

The net heat flow between the two adjacent adiabatic lines can be estimated as

$$\Delta Q \approx k \Delta s Z \frac{\Delta T}{\Delta n} \tag{2.33}$$

where Δs is the distance between adjacent heat flow lines and Δn the distance between isotherms at T and $T + \Delta T$, as shown in Figure 2.15.

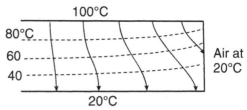

Figure 2.14. Isotherms and adiabatic lines.

For steady conduction in the absence of internal heat generation, ΔQ is fixed for any two adiabatic lines. Thus, following the same two adiabatic lines, the ratio of $\Delta s/\Delta n$ remains constant if the temperature interval between successive isotherms is the same for steady state. For example, if the adiabatic lines and the isotherms roughly form squares near the hot surface, succeeding intersections should also roughly form figures with equal ratios of Δs and Δn. Using this technique, a sketch of the conditions near the side wall of the conductivity tester can be made. Since the air along the side walls is at the same temperature as the lower plate, we should expect there will be some heat transfer from the upper plate to the side wall. Without knowing the convective coefficient along the side wall, only an approximate sketch can be made, but this may still be adequate to get a general picture of the conditions.

Figure 2.14 shows, as we would expect, that the presence of heat transfer to the side wall lowers the temperature in the material near the edge compared to the undisturbed, one-dimensional pattern. Note that because the side wall has a convective boundary condition, the side wall itself is not at 20°C. Care should be taken to draw the adiabatic lines perpendicular to the upper and lower isothermal surfaces. Similarly, the isotherms should be perpendicular to adiabatic surfaces such as the centerline axis of symmetry.

Upper Limit

One upper limit estimate of the heat loss for the sides is to assume the lateral conductivity of the material is high enough that the temperature field remains one-dimensional. That is, along the material's free edge, the temperature variation in the x direction (the vertical direction in Figure 2.14) is the same as its variation near the center ($y = 0$).

Figure 2.15. Isotherms and adiabatic lines.

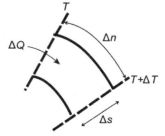

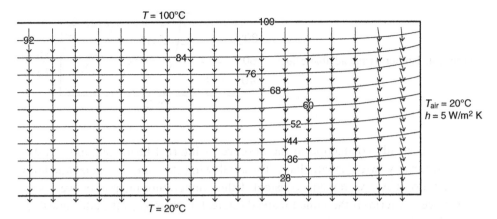

Figure 2.16. Numerical solution for side wall heat transfer.

This gives a temperature along the side that is definitely higher than the real case, resulting in a higher heat loss – an upper limit. In this case, the convective loss for the side becomes

$$Q_{side} \leq Z \int_0^{X_0} h_{sidewall} \, (T(x) - 20) \, dx \tag{2.34}$$

Since T is assumed to vary linearly with x, the integral of $T(x)$ can be replaced by its average, 60°C, and the estimate for Q becomes $40 \, Zh_{sidewall}$. Note that a simpler upper limit can be obtained by assuming that the entire side wall is at the maximum temperature, 100°C. It is worthwhile to explore several upper limits to find the one closest to the correct answer, that is, an upper limit that yields the lowest value.

The magnitude of the upper limit loss can be compared to the one-dimensional heat transfer, with no heat transfer to the side wall

$$Q_{1D} = kYZ\frac{100 - 20}{X} \tag{2.35}$$

where Y is the half width and X the thickness of the sample. The ratio becomes

$$\frac{Q_{side}}{Q_{1D}} \leq \frac{40ZXh_{sidewall}}{80kYZ/X} = \frac{h_{sidewall}X^2}{2kY} \tag{2.36}$$

If the material being tested has a conductivity k similar to that of wood (0.2 W/m K) and we use a value of h for natural convection (5 W/m^2 K), this ratio for a sample 2 cm thick and 1 m wide becomes

$$\frac{Q_{side}}{Q_{1D}} \leq \frac{5 \, (0.02)^2}{2(0.2)(0.5)} = 0.01 \text{ or } 1\% \tag{2.37}$$

Thus, even in the upper limit estimate, the side wall loss is modest.

Note, however, if a material with a conductivity much lower than wood was tested, the upper limit estimate of the error would be considerably larger. On the other hand, if the material had a very low conductivity, the limiting assumption of very large lateral conductivity would be farther from the truth. Figure 2.16 shows

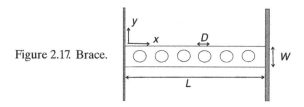

Figure 2.17. Brace.

the results of a two-dimensional numerical solution. For the parameters used with Eq. (2.37), the numerical solution yields an edge loss that is 0.9% of the one-dimensional conduction heat transfer. One-dimensional conditions prevail over the center of the sample away from the sides. Within a distance of two to three sample thicknesses from the sides, two-dimensional effects are observed. As the sample width to thickness increases, the sidewall effects constitute a smaller proportion of the entire area and result in smaller overall errors.

For this case, the upper limit estimate is important because it gives a limit to side loss errors in the test. A lower limit estimate is not as important. In this example, a simple lower limit estimate for the side loss (other than 0) is not obvious.

If the sample under test is not homogeneous, two-dimensional effects may be far more important, as will be shown in a later example.

Example 2.9 Conduction in a structural brace A structural steel brace (Fig. 2.17) extends between the interior and exterior hull of a deep submersible vessel. The space between the walls is air filled to provide insulation. For operation in cold waters, it is desirable to minimize heat transfer along the brace. One suggestion is to space holes on a regular interval along its length, assuming this does not significantly weaken the brace. We need to estimate how much the holes will reduce the heat transfer along the brace.

As a first step, an estimate of the upper and lower limits of the conduction heat transfer along the modified brace will be made assuming no heat transfer to the air-filled cavity.

If the element is rectangular with a thickness b and has fixed temperatures at the two ends, then as a first approximation the heat flow is two dimensional. The approximate isotherms and adiabatic lines for one of the repeating sections may be sketched using the graphical technique of the previous section, as shown in Figure 2.18.

To obtain a lower limit estimate of the heat transfer, we neglect conduction and radiation heat transfer through the circumference of the holes. Further, the

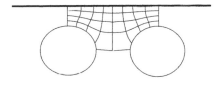

Figure 2.18. Sketch of isotherms and adiabatic lines for one section of the brace.

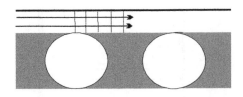

Figure 2.19. Lower limit.

y direction conductivity will be assumed to be negligible for the steel. The heat flow lines, or adiabatic lines, will be parallel to the *x* direction and restricted to the minimum space between the edges of the hole and the outside of the steel as shown on Figure 2.19. The steel between the holes will not conduct any heat.

For an upper limit, consider that the *y* dimension conductivity is very large in the steel brace. In this limit the temperature will be uniform in the *y* direction and the isotherms will all be aligned in that direction; see Figure 2.20. Since the *x* and *y* conductivities are not the same, the adiabatic lines are no longer perpendicular to the isotherms. In this upper limit estimate, the total conduction heat transfer rate at each cross section, in the *x* direction, is the same and *Q* is related to the overall temperature difference by

$$\Delta T = \frac{Q}{k_x} \int \frac{dx}{A_{yz}} \tag{2.38}$$

where A_{yz} is the area of the section in the *y-z* plane at each *x* location.

Example 2.10 *Use of analogies*   One class of insulations is closed-cell foam. In foam, roughly spherical volumes containing gas are surrounded by a continuous solid polymer. Each volume is small enough to eliminate any convection within the gas. The principal form of heat transfer is conduction through the polymer and the gas. If the spherical volumes are not too close together the overall contribution by conduction can be estimated by analogy to direct current electrical conduction. James Clerk Maxwell [2.2] developed a relationship for the electrical conductivity when spheres of one material are suspended in a continuous second material. His relationship for the effective conductivity of both materials taken together is

$$\frac{k_{\text{effective}}}{k_s} = \frac{(2\delta + 1) + 2(1 - \delta)(k_c/k_s)}{(k_s/k_c)(1 - \delta) + (2 + \delta)} \tag{2.39}$$

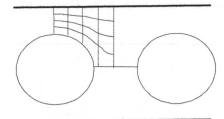

Figure 2.20. Upper limit.

where k_s and k_c are the conductivities of the spheres and the surrounding continuous material, respectively, and δ is the volume fraction of the spheres.

For the limiting case where the conductivity of the spheres (the gas in this case) is much smaller than the polymer continuous material, $k_c/k_s \gg 1$, Eq. (2.39) approaches a simpler expression:

$$k_{\text{effective}} \rightarrow \frac{2(1-\delta)k_c}{(2+\delta)}$$

When the volume fraction of the spheres, δ, is $1/5$, $k_{\text{effective}}$ is $0.72\,k_c$. Inclusion of low-conductivity spheres, even spheres enclosing a vacuum, have only a modest influence on the overall conductivity of the medium. Note, Maxwell's solution holds when the spheres are widely separated, so it should not hold as δ approaches unity.

2.6 Fins

For certain geometries it is possible to simplify the approach to two-dimensional convection heat transfer. If the body is slender compared to its length the temperature change across the short dimension can sometimes be neglected. Fins are the most common example. Consider the slender fin shown in Figure 2.21. One end is attached to a large solid body while the upper and lower surfaces are cooled by convection heat transfer. The third dimension Z is much larger than the fin thickness W so that heat transfer will be considered in only the x-y plane.

To estimate the importance of conduction resistance across the fin thickness, it will be compared to the convection resistance at the surface. Assuming both the top and bottom surfaces of the fin are cooled equally, then the maximum y dimension conduction resistance corresponds to a length $W/2$. This is equivalent to assuming all of the heat must be transferred from the centerline of the fin to the top and bottom surfaces, a limiting condition. For an element of length dx the ratio of the y direction conduction resistance and the corresponding convection resistance with a surface area $Z\,dx$ is

$$\frac{R_{Y\,\text{conduction}}}{R_{\text{convection}}} \leq \frac{\dfrac{W/2}{kZ\,dx}}{1/hZ\,dx} = \frac{hW/2}{k} \tag{2.40}$$

When $R_{Y\,\text{conduction}}$ is an order of magnitude less than $R_{\text{convection}}$, the former can be neglected. In other words when hW/k, the Biot number, is less than about $1/5$ the

Figure 2.21. Two-dimensional fin.

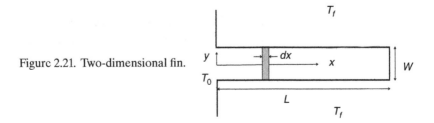

conduction resistance across the width can be neglected and the fin can be assumed to have a uniform temperature across its width. Under these conditions only the conduction in the x direction and the convection at the surface need be considered.

Writing an energy balance over the slice of length dx in the fin for steady-state conditions, we obtain

$$kA\frac{d^2T}{dx^2} - hP(T - T_f) = 0 \tag{2.41}$$

where A is the cross-sectional area normal to the x-axis and P is the wetted perimeter, in this case $2Z\,dx$. At the wall, $x = 0$, the fin temperature will be taken as T_0. This provides one boundary condition for Eq. (2.41). It is useful to look for the solution that embodies the fundamental physics while requiring the simplest level of mathematics. Such a solution serves as a simple check for the more detailed cases to come. In this case if the fin is very long, the temperature at the end should approach the fluid temperature T_f.

To make Eq. (2.41) homogeneous, the variable T will be replaced by φ, defined as $T - T_f$. Equation (2.41) and its two boundary conditions can be rewritten as

$$\frac{d^2\varphi}{dx^2} - \frac{hP}{kA}\varphi = 0 \tag{2.42}$$

$$\text{at } x = 0, \ \varphi = T_0 - T_f; \text{ at } x \to \infty, \varphi \to 0$$

The solution for φ is

$$\varphi = Ae^{mx} + Be^{-mx} \tag{2.43}$$

By substituting this into Eq. (2.42), m is found to be $\sqrt{\frac{hP}{kA}}$.

The boundary condition at large x requires A to go to 0 and the condition at $x = 0$ requires B to be equal to $T_0 - T_f$. The solution for φ is then

$$\varphi = (T_0 - T_f)e^{-\sqrt{\frac{hP}{kA}}x} \tag{2.44}$$

and the heat transfer rate from the base to the surrounding fluid through the fin becomes

$$Q = -kA\left(\frac{dT}{dx}\right)_{x=0} = -kA\left(\frac{d\varphi}{dx}\right)_{x=0} = \sqrt{hPkA}(T_0 - T_f) \quad \text{for large } L \tag{2.45}$$

For the case of a finite length fin with an insulated tip, the corresponding solution is

$$Q = \sqrt{hPkA}(T_0 - T_f)\tanh\left(L\sqrt{\frac{hP}{kA}}\right) \quad \text{for } \frac{dT}{dx} = 0 \text{ at } x = L \tag{2.46}$$

For long fins the two solutions coincide. The fin efficiency, the ratio of actual heat transfer to that for an isothermal fin having at the same temperature as the wall

$$\eta = \frac{Q}{Q(\text{ideal } k \to \infty)} = \frac{\tanh\sqrt{\frac{hPL^2}{kA}}}{\sqrt{\frac{hPL^2}{kA}}} \tag{2.47}$$

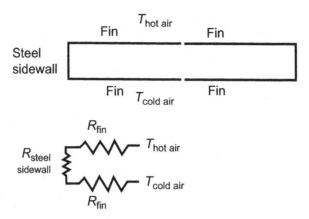

Figure 2.22. Vacuum insulation.

By ingenious modeling, the fin solution can be applied to a wide range of situations that do not, at first glance, look like fins. The next example illustrates this.

Example 2.11 Vacuum insulation In Example 2.7, the vacuum insulation panel, the upper and lower limiting cases using simple resistors in series and in parallel gave widely differing results. This problem will now be revisited to determine whether lateral conduction through the metallic envelope on the two large faces (Figure 2.8) has a first-order influence on the overall insulation performance. To simplify the estimate, an insulation panel where the dimension into the page is long will be considered, so that the problem is two dimensional. To obtain a first lower limit estimate of the influence of envelope conduction on the performance, the evacuated powder will be assumed to be a perfect insulator, that is, that there is negligible heat transfer through the powder. The simplified geometry is shown in Figure 2.22. By symmetry only one-half of the envelope needs to be considered. The equivalent resistance circuit is also shown on the figure.

Using the same dimensions as in Example 2.7, the thermal resistance for one leg of the fin can be calculated noting that L is one half the panel width, or ¼ m. This yields

$$R_{fin} = \frac{1}{\sqrt{hPkA}\,\tanh L\sqrt{\dfrac{hP}{kA}}} \approx \frac{1}{\sqrt{5Z \cdot 20 \cdot 10^{-4}Z}\,\tanh \dfrac{1}{4}\sqrt{\dfrac{5Z}{20 \cdot 10^{-4}Z}}}$$

$$\approx \frac{10}{Z}\,(\text{W/m K})^{-1}$$

$$R_{steelside} \approx \frac{L}{kA} \approx \frac{2 \cdot 10^{-2}}{20 \cdot 10^{-4}Z} = \frac{10}{Z}\,(\text{W/m K})^{-1} \qquad (2.48)$$

The fin resistance includes the lateral conduction along the steel forming the upper surface as well as the convection from the surface to the surrounding fluid. The total thermal resistance for each half, two fins plus the steel sidewall in series, becomes 30/Z. The two half sides are in parallel so that the overall resistance is

15/Z. In the actual, three-dimensional case with a square panel and four steel sidewalls the overall resistance of the envelope will be reduced further. To put this result in perspective, consider the resistance of the powder, which has an effective conductivity of 0.5×10^{-2} W/m K, with the same convective heat transfer on both the top and bottom surfaces

$$R_{powder} + R_{convective} \approx \frac{2 \cdot 10^{-2}}{0.5 \cdot 10^{-2} \cdot 0.5Z} + \frac{2}{5 \cdot 0.5Z} \approx \frac{9}{Z} \text{ (W/m K)}^{-1} \qquad (2.49)$$

The thermal resistances for the two limiting cases, the envelope alone and the powder alone, are almost equal, a clear indication that conduction along the steel envelope will cause a substantial degradation of the overall panel performance.

Example 2.12 *Upper and lower limits to performance of a "fat" fin* Equation (2.40) established a clear criterion for using the fin approximation. What happens when a slender body looks like a fin but the convective heat transfer coefficient is so large that the criterion is not satisfied? Similarly, we could consider the case of a thick stubby fin – the conduction resistance across the fin thickness now becomes important. One recourse in these situations is to treat the body using a full two-dimensional approach. However, the fin solution is still useful. In this case it represents an upper limit for the actual heat transfer through the fin because the fin solution ignores one component of thermal resistance, the conduction resistance across the fin thickness. A lower limit for the actual heat transfer can be obtained by maximizing the possible conduction resistance across the fin thickness, normal to the fin length, and putting it in series with the convection resistance at the surface. If the fin is cooled on both the upper and lower surface, the maximum distance will be from the centerline to the surface. The fin solution is again relevant, but in this case an effective convective resistance is used

$$\frac{1}{h_{effective}} = \frac{1}{h} + \frac{w/2}{k} \qquad (2.50)$$

Equation (2.50) applies to the case when the fin is convectively cooled on both sides. If only one side is cooled, the conduction resistance in Eq. (2.50) should be w/k. Figure 2.23 shows the upper and lower limiting solutions for fin efficiency η compared to a two-dimensional numerical solution when the fin criterion does not hold. In Figure 2.23, the fin length to thickness ratio, L/W, was held constant at a value of 5 as the Biot number was varied. When the Biot number decreases, the two dimensional solution converges to the upper limit, the classical fin case.

2.7 Nondimensional Form of the Governing Differential Equations

When the governing differential equation and boundary conditions can be written for a process, the magnitude of the controlling parameters can sometimes be estimated without solving the equation. The equation is put in nondimensional form and

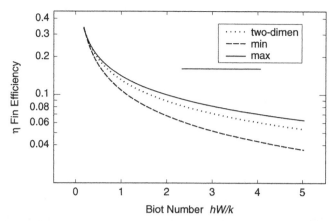

Figure 2.23. Fin efficiency: two-dimensional numerical solution, upper and lower limits, $L/W = 5$ for all cases.

the magnitudes of the terms are assessed. The general rules can be stated, following Kline [2.3]:

1. Nondimensionalize each variable by dividing it by an appropriate characteristic parameter so that the dimensionless variable will vary between roughly 0 and 1. Alternatively, the incremental change of the variables over the domain of interest should be of order unity. The variables consist of the physical quantities of interest that vary over the domain, for example, temperature and velocity fields, spatial coordinates, and time. Given parameters are composed from the boundary conditions, from the characteristic scales of the problem, such as the overall length and temperature at a boundary or the initial velocity, and physical properties. Some care must go into the selection of the appropriate characteristic reference parameter. For example, in a transient process the appropriate time reference may not be obvious.
2. Write the differential equations, boundary conditions, and initial condition in terms of the nondimensional variables.
3. For each equation, divide every term in the equation by the coefficient of one term in that equation to make the entire equation dimensionless. Sometimes, the choice of which coefficient is used to divide the equation is critical. Do the same for the boundary and initial conditions.

The resulting dimensionless coefficients of the terms in the equations represent the set of parameters governing the process. That is, the dimensionless variables are functions of the resulting dimensionless coefficients. The dimensionless form of the equations may also suggest scaling models and analogies to the case in hand. The resulting dimensionless equation can be used to estimate the relative importance of the various terms. This is best illustrated by use of an example.

Example 2.13 *Heat transfer during glass fiber spinning* Consider a case where conduction is evaluated in the presence of convection and through-flow from a single strand of glass being spun into a small diameter fiber. This is a continuation

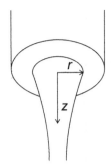

Figure 2.24. Glass fiber spinning.

of the example given in Chapter 1 (see Examples 1.2 and 1.5). Figure 2.24 illustrates the process. High-viscosity liquid glass leaves the nozzle and is attenuated by the tension supplied by the winder at the bottom. After the glass has left the nozzle and moved down a distance equal to twice the nozzle internal diameter, radiation heat transfer within the glass becomes negligible [2.4]. The glass diameter varies from about 0.1 mm down to about 12 μm. The glass temperature ranges from about 1000°C to near ambient. Solutions for the case of a slender long cylinder moving axially in still air suggest that the convective heat transfer coefficient can be estimated as Ck_{air}/D, where D is the local fiber diameter and C may range from unity to more than 10. For these small diameters, this estimate yields a convective heat transfer coefficient of 600 W/m K or more. Even if the glass and the surroundings are considered black bodies, radiation heat transfer as given by Table 2.2 is an order of magnitude less than convection and may be ignored. At the glass surface the thermal boundary condition becomes

$$-k_{glass}\frac{\partial T}{\partial r} = h(T - T_{air}) \tag{2.51}$$

The temperature may be nondimensionalized as

$$\theta = \frac{T - T_{air}}{T_0 - T_{air}} \tag{2.52}$$

and the radius may be nondimensionalized in terms of the glass diameter as

$$\bar{r} = \frac{r}{D} \quad \text{and} \quad \bar{z} = \frac{z}{L} \tag{2.53}$$

where T_0 represents the temperature of the glass at the exit of the nozzle, $Z = 0$, L is the vertical length of the spinning process, and D is the glass diameter. In this way, the nondimensional temperature varies from unity (one) near the glass exit to a value close to 0 near the bottom while $\bar{z}$ varies from 0 to 1. Equation (2.51) can be rewritten as

$$k_{glass}\frac{T_0 - T_{air}}{D}\frac{\partial\theta}{\partial\bar{r}} = h(T_0 - T_{air})\theta \tag{2.54}$$

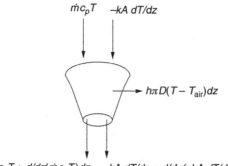

$\dot{m}c_pT \qquad -kA\ dT/dz$

$\rightarrow h\pi D(T - T_{air})dz$

$\dot{m}c_pT + d/dz(\dot{m}c_pT)dz \qquad -kA\ dT/dz + d/dz(-kA\ dT/dz)dz$

Figure 2.25. Glass cross-section.

Writing Eq. (2.54) in nondimensional form and simplifying it, we can estimate the magnitude of the temperature difference across the glass cross section

$$\frac{\partial \theta}{\partial \bar{r}} = \frac{hD}{k_{glass}}\theta$$

$$\frac{\Delta \theta}{\Delta \bar{r}} \approx \frac{k_{air}}{D}\frac{D}{k_{glass}}\theta \ll 1 \tag{2.55}$$

The left-hand side can be estimated as a change in θ from the centerline to the outside surface, as $\bar{r}$ changes from 0 to 1/2. Since the heat transfer coefficient is roughly k_{air}/D then the right-hand side is proportional to the ratio of air to glass conductivity, a magnitude much less than unity. Thus, the radial temperature difference across the fiber can be neglected and the glass can be assumed to have a one-dimensional temperature distribution, varying only with z.

To obtain the variation of glass temperature with axial distance, an energy balance is written for a control volume over the glass cross section shown in Figure 2.25. The energy balance for steady-state flow becomes

$$-\frac{\partial}{\partial z}\left(k_{glass}A\frac{\partial T}{\partial z}\right) + \dot{m}c_p\frac{\partial T}{\partial z} + h\pi D(T - T_{air}) = 0 \tag{2.56}$$

with the boundary conditions

$$\text{at } z = 0,\ T = T_0 \quad \text{and} \quad \text{at } z \to L,\ T \to T_{air}$$

Nondimensionalizing T and z using Eqs. (2.52) and (2.53), respectively, Eq. (2.56) becomes

$$\frac{k_{glass}A}{L^2}\frac{\partial^2\theta}{\partial \bar{z}^2} + \dot{m}c_p\frac{(T_0 - T_{air})}{L}\frac{\partial \theta}{\partial \bar{z}} + h\pi D(T_0 - T_{air})\theta = 0 \tag{2.57}$$

Dividing each term in the equation by the coefficient of the first term results in

$$\frac{\partial^2\theta}{\partial \bar{z}^2} + \frac{\dot{m}c_pL}{k_{glass}A}\frac{\partial \theta}{\partial \bar{z}} + \frac{h\pi DL^2}{k_{glass}A}\theta = 0 \tag{2.58}$$

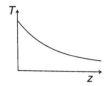

Figure 2.26. Monotonic variation.

If the variation of θ with $\bar{z}$ is smooth and monotonic as shown on Figure 2.26, the first derivative of θ with $\bar{z}$ can be estimated as

$$\frac{\partial \theta}{\partial \bar{z}} \approx \frac{\theta_{z=1} - \theta_{z=0}}{\Delta \bar{z}} \approx \frac{0-1}{1} \approx O\,[1] \tag{2.59}$$

And the second derivative of θ with $\bar{z}$ can be estimated as

$$\frac{\partial^2 \theta}{\partial \bar{z}^2} \approx \frac{\left.\frac{\partial \theta}{\partial \bar{z}}\right|_{z=1} - \left.\frac{\partial \theta}{\partial \bar{z}}\right|_{z=0}}{\Delta \bar{z}} \approx \frac{0-1}{1} \approx O\,[1] \tag{2.60}$$

where, for $\bar{z}$ approaching 1, the value of $\frac{\partial \theta}{\partial \bar{z}}$ should approach 0. Thus, the order of magnitude of the three terms in the governing differential equation becomes

$$\underbrace{\frac{\partial^2 \theta}{\partial \bar{z}^2}}_{O(1)} + \left(\frac{\dot{m}c_p L}{k_{glass}A}\right)\underbrace{\frac{\partial \theta}{\partial \bar{z}}}_{O(1)} + \left(\frac{h\pi DL^2}{k_{glass}A}\right)\underbrace{\theta}_{O(1)} = 0 \tag{2.61}$$

Physically, the last term in the left-hand side represents convective cooling of the fiber. If this term were negligible there is no cooling and a solid fiber cannot be formed. This third term will therefore always be of a size similar to either the first term or the second term. When the coefficient of the second term, representing the flow of enthalpy of the glass fiber, $\frac{\dot{m}c_p L}{k_{glass}A}$, is much greater than unity, the first term, representing axial conduction in the fiber, can be neglected. This second term coefficient can be recast as

$$\frac{\dot{m}c_p L}{k_{glass}A} = \frac{\rho U A c_p L}{kA} = \frac{UL}{\alpha} \quad \text{the Péclet number} \tag{2.62}$$

where $\alpha = k/\rho c_p$ is the thermal diffusivity. If, on the other hand, the Péclet number is much smaller than unity, energy transfer due to glass flow is small compared to axial conduction and the heat transfer process approaches that of a stationary fin, with a balance of axial conduction through the glass fiber and convective cooling at its surface.

When the Péclet number is large and the conduction term can be neglected, then the governing equation can be reduced to the two last terms of the left-hand side. Note in this case, there are not two independent dimensionless parameters controlling the process. The two coefficients can be combined into one, recasting the equation as

$$\frac{\partial \theta}{\partial \bar{z}} + \frac{h\pi DL}{\dot{m}c_p}\theta = 0 \tag{2.63}$$

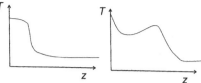

Figure 2.27. "Non-regular" temperature variation.

In a situation where the variation of temperature with z is not as regular as that shown in Figure 2.26 (see Figure 2.27), the estimation of the derivatives is not as straightforward and different estimates would be required. For example, the appropriate reference length for nondimensionalizing z is not necessarily the total length L in such cases.

2.8 Use of Known Solutions: Conduction Shape Factors

For steady-state conduction between two isothermal surfaces separated by a medium of uniform conductivity, the form of the heat transfer rate for a particular configuration is

$$Q = k(\Delta T) f\left(\frac{A}{w}\right) \tag{2.64}$$

where f is a function of some mean surface area separating the two surfaces to a mean thickness between the two. In more general cases, the heat transfer rate can be rewritten in terms of a shape factor S that has the units of length

$$Q = kS\Delta T \tag{2.65}$$

Values of S have been found by analytical and numerical means for a variety of shapes. Table 2.3 gives some representative values. It must be kept in mind that the values given apply only when prescribed conditions hold: steady state conduction through a homogeneous medium with isothermal bounding surfaces. Additional conditions apply for some shape factors. For example, for the buried cylinder, the soil far below the ground level must be at the same temperature as the soil surface.

As before, we can develop some approximate upper and lower limiting solutions if the shape factor is not available for the geometry in question. As an example, consider conduction from a long isothermal body of square cross section to a concentric surrounding body that is also isothermal (Figure 2.28, left side).

A lower limit for the heat transfer can be found by using the surface area of the inner body and the wall thickness to calculate

$$Q_{\text{MIN}} = \frac{k4bZ(\Delta T)}{w} \tag{2.66}$$

while an upper limit can be found using the surface area of the outer body in the calculation

$$Q_{\text{MAX}} = \frac{k4aL(\Delta T)}{w} = \frac{k4(b+2w)L(\Delta T)}{w} = \left(1 + \frac{2w}{b}\right) Q_{\text{MIN}} \tag{2.67}$$

Table 2.3. *Some representative shape factors for steady-state conduction with isothermal boundaries*

Long concentric cylinders with radii r_2 and r_1 and length L	$\dfrac{2\pi L}{\ln(r_2/r_1)}$ for $L \gg r_2$
Long concentric bodies with square cross sections 	$\dfrac{2\pi L}{0.93\ln(a/b) - 0.0502}$ for $a/b > 1.4$ $\dfrac{2\pi L}{0.785(a/b)}$ for $a/b < 1.4$
Buried cylinder 	$\dfrac{2\pi L}{\cosh^{-1}(z/r)}$ $\dfrac{2\pi L}{\ln(2z/r)}$ for $z > 3r$
Conduction from an isothermal spherical hole of radius R to an infinite medium 	$4\pi R$
Buried sphere Medium at infinity also at temperature of surface	$\dfrac{4\pi R_1}{1 - \dfrac{R_1}{2h}}$
Long parallel cylinders exchanging heat in an infinite conducting medium 	$\dfrac{2\pi}{\cosh^{-1}\left(\dfrac{L^2 - R_1^2 - R_2^2}{2R_1 R_2}\right)}$

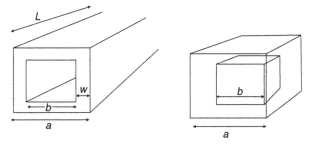

Figure 2.28. Long isothermal square bodies and centered isothermal cubes.

When the wall thickness is 10% of the inner square dimensions then the maximum and minimum solutions differ by 20% and a solution that used the area half way between the inner and outer perimeters would differ from the exact solution by at most $\pm 10\%$. A solution using the approximation

$$Q \approx \frac{k4 \left(\dfrac{a+b}{2} \right) L(\Delta T)}{w} \tag{2.68}$$

differs from the shape factor in Table 2.3 by less than 3% when a/b is 1.2 and less than 14 percent when a/b is 2.

Life is not so simple for conduction in the medium between an isothermal cubic surface centered inside of another cube (Figure 2.28, right side). The simple approximation using the area of a cubic surface centered between the inside and outside surface along with the wall-to-wall spacing gives much larger errors. This should be expected because the lower and upper limit solutions using the inner and outer surface areas, respectively, differ by 45% when the width of the outside cube is 20% larger than the inside cube. Note that the correct results can be found by use of a combination of shape factors given in Langmuir et al. [2.5]. Numerous analytical solution techniques are discussed by Carslaw and Jaeger [2.6].

REFERENCES

[2.1] W. M. Rohsenow and H. Y. Choi. *Heat Mass and Momentum Transfer*. Englewood Cliffs, NJ: Prentice Hall, 1961.
[2.2] J. C. Maxwell. *Electricity and Magnetism*. Oxford: Clarendon Press, 1873.
[2.3] S. Kline. *Similitude and Approximation Theory*. New York: McGraw-Hill, 1965.
[2.4] L. R. Glicksman. The dynamics of a heated free jet of variable viscosity liquid at low Reynolds numbers. *Journal of Basic Engineering, Transactions of the ASME*, **90**(3), 343–354, 1968.
[2.5] I. Langmuir, E. Q. Adams, and G. S. Meikle. Flow of heat through furnace walls. The shape factor. *Transactions of the American Electrochemical Society*, **14**, 53–81, 1913.
[2.6] H. Carslaw and J. Jaeger. *Conduction of Heat in Solids*, 2nd ed. Oxford: Oxford University Press, 1959.

PROBLEMS

2.1 One proposal for an advanced thermal insulation to be used inside refrigerator walls is the flat evacuated panel shown in Figure 2.29. The upper and lower surfaces are thin metal sheets that are welded together at the edges to form an enclosed space that is kept under vacuum conditions. The metal sheets have a thickness t and a conductivity k. Glass spheres arranged in a widely spaced square pattern are used to support the metal sheets against the pressure difference. The inside of both metal sheets has a low emissivity, ε.

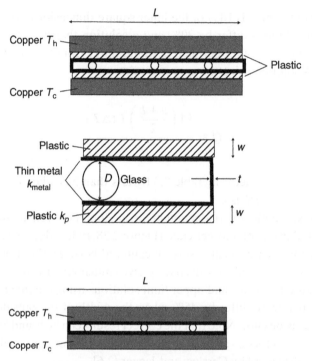

Figure 2.29. Two measurement methods.

The R factor, R_f, also referred to as the R value, is the ratio of the overall temperature difference across the insulation panel and the measured heat flux per unit area.

Two methods are proposed to measure the overall R value of the panel. In the first method the panel is sandwiched between two constant temperature copper blocks with a low-conductivity plastic sheet of thickness w and conductivity k_p between each metal sheet and copper block. In the second method, the metal sheets are in direct contact with the constant temperature blocks. The R value with the plastic sheets is the measured overall R value of the insulation plus plastic sheets less the one-dimensional R value of the two plastic sheets. Will both methods give the same value of R_f?

Look at upper and lower limit cases to help determine if both methods give the same value of R value for the thermal insulation.

Figure 2.29 shows both methods of measuring R_f.

2.2 Closed-cell foam insulation such as polyurethane has a low-conductivity gas contained in closed cells where walls are formed by a solid polymer. The cells are isotropic (sometimes) with a diameter of approximately 0.2 mm, thus excluding any significant heat transfer by natural convection. Typically, 3% of the volume is occupied by solid and the balance by the gas. We need to estimate the net conduction heat transfer through the solid and gas. Do not include radiation. Consider the idealized array of cubic cells shown in Figure 2.30, which are offset in one dimension (normal to the applied temperature gradient). The thickness of the cell walls is uniform. Obtain an upper and lower limit estimate of the combined heat flux as a means of bracketing the true value. The plastic conductivity is 0.2 W/m°C and the gas conductivity is 0.01 W/m°C. [At a minimum, develop the electrical analogy for these two limits.]

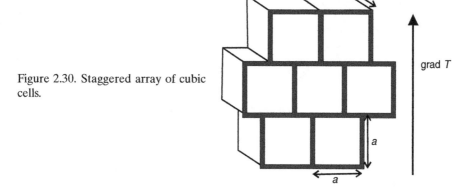

Figure 2.30. Staggered array of cubic cells.

Note the cells touch all adjacent cells (there isn't any gap between them)

2.3 Steel members are increasingly being used as structural elements in home construction. Figure 2.31 shows a steel framing system for one half of a roof. The steel rafters are covered by wood sheathing to which conventional shingles are nailed. Gypsum wallboard is attached to the bottom of the steel joists and the space between the joists is filled with conventional fiberglass insulation. The vertical wall framing consists of two by four channels. The attachment between rafters and joists is secure or so that there is negligible contact resistance. The attic is vented to allow possible external airflow through it. A major concern is the overall heat loss from the interior ceiling through the roof system. Bracket the correct value of heat loss by using maximum and minimum estimates combined with suitable heat transfer models.

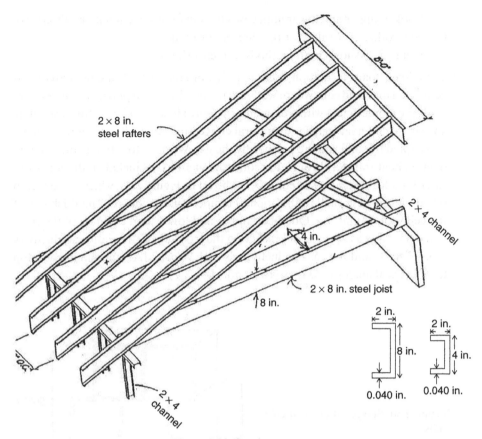

Figure 2.31. Steel structure.

2.4 Problem 2.1 describes a vacuum panel made with metal sheet. Consider how to use the fin solution to achieve a lower limit estimate of the influence of conduction along the thin metal sheet as it wraps around the circumference of the panel, from the hot to the cold side. Assume the panel is installed within a refrigerator and is in contact with air on both of its flat surfaces. Set up the estimate but don't solve.

2.5 To minimize the edge loss in a vacuum insulation panel, the stainless steel sheet that encloses the evacuated space is folded several times at the edge as shown in Figure 2.32. Insulation is inserted between each of the folds. The top and bottom are exposed to air at temperatures of T_h and T_c, respectively with convective heat transfer coefficient h. Neglect radiation heat transfer from the surface. The width of the folded region is W and the thickness of each fold is L. The length of the edge layer in the third dimension is Z. Stainless steel has a thickness of t and conductivity k_s. The insulation material has a conductivity k_i. Consider only the two-dimensional heat transfer through the edge layer. Neglect the lateral heat transfer between the edge region and the main body of the insulation.

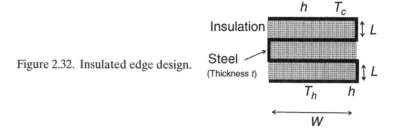

Figure 2.32. Insulated edge design.

Make a reasonable upper limit estimate of the total heat transfer through the edge region. Show the corresponding electrical analogy labeling all of the resistance elements. Clearly state your assumptions.

2.6 A large fin is designed with a tapered cross section so that the thickness t (in the y direction) varies with the distance x from the base. The fin cross section remains the same in the dimension z normal to the fin profile. The fin is made of aluminum and has a uniform convective heat transfer coefficient of 20 W/m² K over its surface. At its thickest, where $t(x)$ is largest, the fin has a Biot number of 0.1.

To test the fin, an engineer proposed to test a smaller sized model that has all linear dimensions one-fifth those of the original large fin. She also proposes to have the same convective heat transfer coefficient, 20 W/m² K over the surface of the model. Neglect radiation.

a. Write down, but don't solve, the governing differential equation and use it to find the governing dimensionless parameters that establish the temperature distribution along the fin.

b. What material must the model fin be made of to simulate the performance of the large fin exactly, that is, to have the corresponding temperature variation in x and y?

c. How does part (a) change if there is radiation from the fin surface to black body surroundings at a temperature T_s that differs from both the fluid temperature T_f around the fin and the temperature at the base of the fin T_b? Assume you can use the linearized form of the radiation heat transfer coefficient, h_r.

2.7 To prevent burns, a handle for a high-temperature machine is made of a coiled metal bar with a rectangular cross section of dimensions W by t. The bar is composed of N helical turns of radius D, as shown in Figure 2.33. One end of the bar is attached to the hot device at temperature T_h. The bar is immersed in air at a cooler temperature T_c. The air space between successive turns is δ which is of the same order as the bar width W.

a. Develop an upper and lower limit estimate of the temperature at the outer end of the bar in terms of T_h, T_c, the bar geometry, conductivity k, and air side heat transfer coefficient h. Be careful to define the area associated with the coefficient h and the influence of the size of the spacing δ.

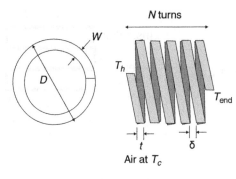

Figure 2.33. Coiled handle.

b. Now suppose that the coiled bar is hollow and is made of a low-conductivity polymer that has an effective conductivity of the same order as air. In addition, the spacing between turns, δ, and the bar thickness t are approximately the same and are both very small. Obtain an estimate for the temperature of the outer end of the bar.

2.8 An aluminum support is 50 cm long and 5 mm thick. It tapers from one end to the other from a maximum height of 2 cm to a minimum height of 1.5 cm, as shown in Figure 2.34. To reduce its weight, 7.5 mm diameter holes are drilled through it along the length with a center-to-center spacing of 10 mm. The end with the 2 cm height is at 60°C, while the other end is at 55°C. The bar is at steady state. It is surrounded on all sides by air at 20°C. The surface heat transfer coefficient is 30 W/m K. The insides of the holes are adiabatic. The aluminum is an alloy having a thermal conductivity of 170 W/m K.

  Make reasonable upper and lower limit estimates of the total heat transfer rate from both ends to the air. A numerical answer is required. Be careful to state your assumptions.

  Indicate how you could improve the upper and lower limit results, that is, how you could bracket the true value more closely.

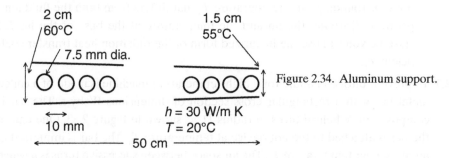

Figure 2.34. Aluminum support.

2.9 Small hollow glass spheres have been produced with a hard vacuum inside them. The interior surface of the glass is coated with a very highly reflecting material to minimize thermal radiation exchange.

    a. If the spheres are embedded in a polymeric medium, estimate how much the effective conductivity is decreased as a function of the volumetric concentration of the spheres.

    b. If the spheres are fused together to form a solid matrix without any polymer, how will this compare to the performance of closed-cell foam that has the same volume fraction of solid material?

2.10 Two cylindrical rollers of diameter D are used in a press. They are forced together so that they deform at the contact point (see Figure 2.35). A region of tight contact between them has a width of $D/10$. The temperature of the upper cylinder is uniform at its centerline, equal to T_1, while the lower cylinder has a cooler centerline temperature of T_2. The exterior surfaces of the cylinder have negligible heat transfer. Sketch the isotherms and heat flow lines between T_1 and T_2. Compare this to the heat flow lines for an upper and lower limit approximation of the heat transfer.

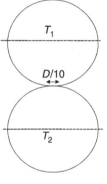

Figure 2.35. Roller press.

3

Transient Conduction Modeling

3.1 Introduction

Our aim in this chapter is to study engineering models for unsteady heating and cooling processes. These models are used to estimate rates of temperature change, heat fluxes, and transient temperature distributions.

In the steady conduction problems examined in the previous chapter, thermal resistance or thermal conductivity was the physical variable of greatest importance. In unsteady heat transfer processes, the heat capacitance of the objects involved must also be considered. Modeling unsteady heat conduction hinges on recognizing the heat capacitances and thermal resistances present.

In some situations, a thermal resistance external to a solid object is much larger than the internal conduction resistance of the object. In such cases, it is possible to neatly separate the main thermal resistance from the main thermal capacitance; such *lumped capacity* situations are characterized by a small value of the Biot number.

In other circumstances, the conduction resistance within a body may be as large or larger than that outside the body (Biot number of order 1 or greater). In these cases, the thermal resistance of a body is distributed throughout, as is the thermal capacitance, and it is usually necessary to solve a partial differential equation to determine the precise temperature distribution within the body. Even so, ballpark estimates of the temperature response can be obtained by using an appropriate approximation of the body's internal resistance (see Section 3.2).

In this chapter, we first examine simple resistance–capacitance models for unsteady thermal response. These estimates extend the usual lumped capacity model. Then, we consider adaptations of analytical conduction solutions for the response of bodies at large time, using series and chart results. Short time models, for semi-infinite behavior, are discussed next. Scaling analysis of conduction problems is introduced in the following section; such analyses are used here mainly as a tool for simplifying complex situations. These ideas are explored further in the end-of-chapter problems.

3.1.1 Physical Properties in Unsteady Conduction

The thermal energy storage of an object, per unit weight and per unit temperature change, is measured using its specific heat capacity, J/kg·K, when no phase change occurs. As you'll recall from thermodynamics, the specific heat capacity during a constant volume process, c_v, is different than that for a constant pressure process, c_p, *unless* the material involved is incompressible. For nearly incompressible materials, such as most solids and some liquids (water in particular), the heat capacity is nearly the same for constant volume and constant pressure situations. In what follows, we are usually considering constant pressure processes, and we will write c_p for both compressible and incompressible materials.[1]

For solid materials, we are often concerned not with the heat capacity per unit weight but with the heat capacity per unit volume, ρc_p, because the internal energy, U, of a body is proportional to $mc_p = \rho c_p \mathcal{V}$, for m the mass, ρ the density, and $\mathcal{V}$ the volume of the body. For nonporous solids, the value of ρc_p does not vary much: it typically ranges from 2×10^6 to 4×10^6 J/m^3 K. The reason for this is that the atoms in all solids have nearly the same spacing – within about 30% of 0.24 nm. These atoms store energy in three vibrational degrees of freedom, amounting to $3k_B T$ per atom, where k_B is Boltzmann's constant, 1.380×10^{-23} J/K. Consequently, the energy stored by solids per unit volume per unit temperature change doesn't vary much from

$$\rho c_p \approx 3k_B/(0.24 \times 10^{-9})^3 = 3 \times 10^6 \text{ J/m}^3\text{K}. \tag{3.1}$$

The near constancy of ρc_p has a profound implication in selecting materials for unsteady heat conduction processes in given volumes, in so far as it shows that only the thermal conductivity can really be varied by choice of material. The one exception to this principle occurs in porous materials, such as insulating foams. For porous materials, having a volume fraction δ, both k and ρc_p vary with δ. We can compute the volumetric heat capacity as

$$\rho c_p = (1 - \delta)(\rho c_p)_{\text{solid}} + \delta(\rho c_p)_{\text{void}} \tag{3.2}$$

(The estimation of k for porous materials was discussed in Chapter 2.) Porous materials may have substantially reduced volumetric heat capacity.

Figure 3.1 is a plot of thermal conductivity k versus thermal diffusivity, $\alpha = k/\rho c_p$, for a spectrum of materials, with lines of constant ρc_p shown. The conductivity ranges over more than four orders of magnitude, and the variation in α matches it along a path of nearly constant ρc_p.

---

[1] For gases, the distinction between constant pressure and constant volume heat transfer matters owing to the $p\,dV$ work that accompanies expansion. In most steady flow convection problems, the pressure does not vary enough in the boundary layer region where heat is transferred to cause changes in thermodynamic properties, and the approximation of a constant pressure process is appropriate. Exceptions occur in unsteady convection processes that are accompanied by large pressure swings, such as are found in reciprocating machinery. Remarkably, when steady compressible flow occurs in a duct, the streamwise pressure gradient is usually low enough that the heat transfer locally can be computed with incompressible models; however, compressibility effects must be taken into account in the streamwise direction to determine the state of the gas in the local cross section.

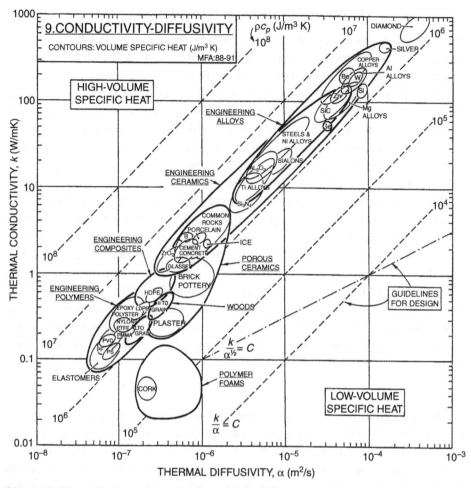

Figure 3.1. Thermal conductivity and thermal diffusivity for a variety of materials (from M.F. Ashby [3.1]).

The graph also shows a line of constant $k/\sqrt{\alpha} = \sqrt{k\rho c_p}$. This quantity appears in semi-infinite body conduction problems (see Table 3.4).

Example 3.1 *Specific heat capacity of glass fiber insulation* In Example 2.5, we considered the thermal conductivity of a glass fiber insulation. The insulation consisted of thin glass fibers surrounded by air, and it had a density of 16 kg/m^3. What is the specific heat capacity of the insulation?

To find this, we can use Eq. (3.2), but first we must compute the volume fraction of glass. The density of the insulation results from both air and glass:

$$\rho_{\text{insulation}} = \delta\rho_{\text{air}} + (1 - \delta)\rho_{\text{glass}} \tag{3.3}$$

With $\rho_{\text{air}} = 1.177$ kg/m^3 and $\rho_{\text{glass}} = 2480$ kg/m^3, we find $\delta = 0.9940$: the insulation is mostly air by volume. Then, with $c_{p,\text{air}} = 1007$ J/kg·K and

$c_{p,\text{glass}} = 750$ J/kg·K, we have:

$$(\rho c_p)_{\text{insulation}} = (1 - \delta)(\rho c_p)_{\text{glass}} + \delta(\rho c_p)_{\text{air}} \tag{3.4a}$$

$$= (1 - 0.9940)(2480)(750) + (0.9940)(1.177)(1007) \tag{3.4b}$$

$$= 12.34 \text{ kJ/m}^3\text{K} \tag{3.4c}$$

On dividing by the density of the insulation, we find $c_{p,\text{insulation}} = 771$ J/kg·K. Most of the specific heat capacity results from the glass, despite the fact that the insulation is mainly air by volume.

When mass is the primary consideration for sensible energy storage, then c_p will be the focus of attention. As may be concluded from the preceding discussion, the value of c_p is generally lower for more dense materials. Values of c_p range widely, from about 130 J/kg·K for lead to about 4180 J/kg·K for water, both near room temperature.

3.2 Thermal Resistance and Thermal Capacitance

The rates at which objects cool are determined by the thermal resistances within the object and at its exterior and the heat capacitance of the objects involved. As a case for discussion, let us consider the cooling of a plate of glass-cloth reinforced epoxy (having the grade name FR-4). This material is commonly used in electronic equipment owing to its high stiffness and low electrical conductivity. In particular, in some electronics testing equipment it is necessary to cycle the temperature of FR-4 structures between an 85°C test temperature and ambient, and quick thermal response is desirable. The thermal properties of FR-4 are: $k = 0.3$ W/m·K, $\rho = 1800$ kg/m^3, and $c_p = 1600$ J/kg·K. It is a relatively poor conductor with volumetric heat capacity similar that of most other nonporous solids.

3.2.1 The Biot Number

The FR-4 plate is to be cooled convectively from both sides. We can start by looking at the thermal resistances. Two must be considered. The first is the resistance of the convective cooling process at the surface, as discussed in Section 2.3. If the surface area is A, then the convective resistance on each side is

$$R_{\text{ext}} = \frac{1}{hA} \tag{3.5}$$

The second resistance is for conduction inside the plate. This resistance requires a little more thought. If we assume the plate thickness is much less than its width and height, and that the cooling is uniform over the surface, then we only need to consider heat transfer through the thickness of the plate. In the initial stages of cooling, only the region near the surface of the plate is involved in cooling, so that heat does not travel far through the plate, and internal conduction resistance for this layer is small. As time progresses, however, the temperature gradient extends all the way to the center of the plate and so the entire half thickness of the plate is involved (the centerline is a line of symmetry across which the flux is 0). If the half-thickness is L, then a good estimate for the average one-dimensional heat conduction resistance for these later stages of cooling (which represent most of the cooling time) might be based on a length of about $L/2$ (or ¼ of the total thickness)

$$R_{int} = \frac{(L/2)}{kA} \tag{3.6}$$

The relative size of these two resistances has a controlling effect on the rate of cooling. We can compare them by forming a ratio and defining the Biot number:

$$\frac{R_{int}}{R_{ext}} = \frac{\bar{h}L}{2k} \tag{3.7}$$

The factor of 2 is not important for order of magnitude considerations, so we will define the Biot number as

$$Bi \equiv \frac{\bar{h}L}{k} \tag{3.8}$$

where its physical significance is essentially unchanged: it is the ratio internal conduction resistance to external convection resistance.

Example 3.2 *Dominant resistance for various convective cooling schemes* Let the FR-4 plate have a half-thickness $L = 5$ mm and suppose that the options for cooling it include letting it sit in still air (natural convection and radiation with a combined value of $\bar{h} = 5$ W/m² K), blowing an array of air jets onto it (forced convection with $\bar{h} = 75$ W/m² K), or cooling it with a flow of fluoroinert liquid (forced convection with $\bar{h} = 600$ W/m² K). What is the dominant resistance in each case?

We may answer this question by computing the Biot number for each case.

Natural convection and radiation:

$$Bi = \frac{(10)(0.005)}{0.3} = 0.16 \ll 1 \qquad \text{Convection resistance dominant} \qquad (3.9a)$$

Forced convection of air:

$$Bi = \frac{(75)(0.005)}{0.3} = 1.3 \qquad \begin{array}{l}\text{Convection and conduction} \\ \text{resistances are similar}\end{array} \qquad (3.9b)$$

Forced convection of liquid:

$$Bi = \frac{(600)(0.005)}{0.3} = 10 \gg 1 \qquad \text{Conduction resistance dominant} \qquad (3.9c)$$

The appropriate model for the cooling rate is different depending on which resistance is largest. In all cases, the heat capacitance of the plate, mc_p, is the same: $\rho c_p(2LA)$ for the entire plate or $\rho c_p(LA)$ for each half of it.

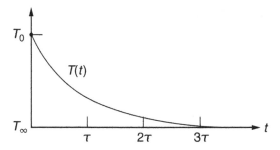

Figure 3.2. Time variation of temperature when external resistance dominates.

External Resistance Dominant: Lumped Capacity Behavior

Now let's look at the case where the plate is cooled by natural convection and radiation in still air. We found that the Biot number was small compared to 1, meaning that the internal conduction resistance is negligible in comparison to the external convection resistance. This also means that the interior of the plate has little temperature gradient and is essentially isothermal. It follows that the internal energy of the plate can be represented by one temperature, T: we can think of it as a thermal capacitance, mc_p, at temperature T. The rate of change of that energy can be related to the rate of heat loss through the surface using the First Law

$$\frac{dU}{dt} = -Q \qquad (3.10a)$$

$$mc_p \frac{dT}{dt} = -\frac{T - T_\infty}{R_{ext}} \qquad (3.10b)$$

where T_∞ is the ambient temperature. This first-order equation has the solution[2]

$$\frac{T - T_\infty}{T_0 - T_\infty} = \exp\left(-\frac{t}{\tau}\right) \qquad (3.11)$$

where the time constant, τ is defined as

$$\tau \equiv mc_p R_{ext} \qquad (3.12)$$

The time variation is as shown in Figure 3.2. The plate will be have undergone 95% of its temperature change after three time constants have elapsed (an interval of 3τ).
We may alternatively write the time constant in terms of the heat transfer coefficient or the Biot number by substituting $1/\bar{h}A$ for the external resistance on one

---

[2] A more general solution for lumped cooling allows the temperature of the environment to vary in time, $T_\infty(t)$:

$$T(t) = T_\infty(t) + [T_0 - T_\infty(0)]e^{-t/\tau} - e^{-t/\tau}\int_0^t e^{s/\tau}\left[\frac{d}{ds}T_\infty(s)\right]ds.$$

side and using $m = \rho LA$ for half the plate mass[3]:

$$\tau = \frac{\rho c_p LA}{\bar{h}A} \tag{3.13}$$

$$= \frac{L^2}{\alpha}\frac{1}{\text{Bi}} \qquad \text{for Bi} \ll 1 \tag{3.14}$$

As the heat transfer coefficient becomes smaller, the time required for cooling increases.

Note that if insulation were placed around the plate, it would add an additional external thermal resistance in series with R_{ext}. If the thermal capacitance of the insulation were negligible in comparison to that of the plate (as is often the case with porous insulations), then the insulation's resistance would simply be added to R_{ext} in the previous equations, since resistances in series sum to give an equivalent thermal resistance.

Internal Resistance not Negligible

Next suppose that we cool the plate from both sides by forced convection with air. We found that Bi = 1.3 in this case, so that R_{int} and R_{ext} are of the same magnitude. At this point, a rigorous calculation would require us to solve the energy equation and to work with the resulting Fourier series. Since our interest at this point is to make a simple *estimate* of the time involved, we'll instead look at an adaption of the lumped capacity solution.

Specifically, we now need to account for the presence of an internal conduction resistance. This resistance is in general a function of time; however, as was previously noted, once the temperature disturbance extends to the center of the plate, the internal resistance remains approximately constant and equal to $L/2kA$. The interior of the plate now has a temperature gradient as well; but let us approximate the internal temperature with a single average temperature, $\bar{T}$, that represents the amount of energy remaining in the plate (we will consider the temperature distribution within the plate later). Specifically,

$$U = \int_0^L \rho c_p[T(x) - T_\infty]A\,dx \approx \rho c_p AL(\bar{T} - T_\infty) \tag{3.15}$$

The effective thermal circuit is as shown in Figure 3.3. The rate of temperature change is again given by the First Law, but with a combined thermal resistance

$$mc_p\frac{d\bar{T}}{dt} = -\frac{\bar{T} - T_\infty}{R_{\text{ext}} + R_{\text{int}}} \tag{3.16}$$

The cooling process is again exponential, and the solution is

$$\frac{\bar{T} - T_\infty}{T_0 - T_\infty} = \exp\left(-\frac{t}{\tau}\right) \tag{3.17}$$

---

[3] Or we could use the equivalent resistance of the parallel resistances for heat loss from both sides, $1/2\bar{h}A$ with the entire plate mass, $2\rho LA$. The same value of τ results.

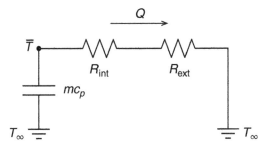

Figure 3.3. Effective thermal circuit when internal resistance is not negligible.

with a new time constant,

$$\tau \equiv mc_p(R_{ext} + R_{int}) \tag{3.18}$$

In terms of the Biot number, the time constant is

$$\tau = \frac{L^2}{\alpha}\left(\frac{1}{\text{Bi}} + \frac{1}{2}\right) \tag{3.19}$$

When the Biot number becomes large, the external resistance is negligible compared to the internal resistance, and the cooling occurs at a rate governed by the conduction resistance of the plate. In this limit, increasing the external heat transfer coefficient does nothing to raise the rate of cooling.

For example, if we cooled our FR-4 plate with forced convection of fluoroinert liquid ($\bar{h} = 600$ W/m^2), Bi = 10 and the time constant would be

$$\tau \approx \frac{1}{2}\frac{L^2}{\alpha} \qquad \text{for Bi} \gg 1 \tag{3.20}$$

Since the time constant is now essentially independent of $\bar{h}$, further efforts to increase $\bar{h}$ would be pointless; likewise, precise prediction of $\bar{h}$ is unnecessary.[4] In this case, 95% of the cooling has occurred after an interval of

$$3\tau = \frac{3}{2}\frac{L^2}{\alpha} \qquad \text{for Bi} \gg 1 \tag{3.21}$$

In each case, we have found that the time constant is proportional to L^2/α. The variation of the nondimensional 95% cooling time, $3\tau\alpha/L^2$, with Biot number is plotted in Figure 3.4. It is useful to note that switching from still air to forced air raises the heat transfer coefficient and the Biot number by about a factor of 8, from Bi = 0.16 to Bi = 1.3, and decreases the cooling time by a factor of 5, from about 1500 s to about 300 s. Switching to fluoroinert cooling, with an additional factor of 8 increase in the heat transfer coefficient (Bi = 10), achieves less than a factor of 3 decrease in the cooling time, to about 120 s, while the increasing the cooling system cost by roughly two orders of magnitude.

---

[4] In fact, for large Bi, the surface temperature is approximately T_∞.

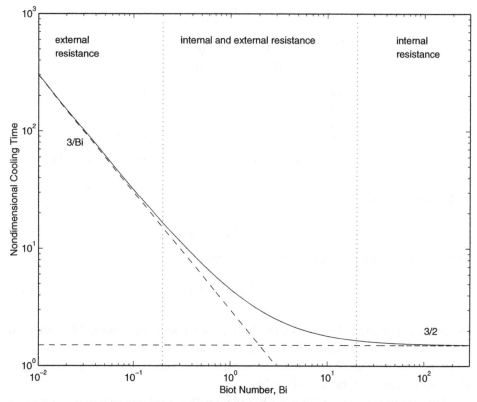

Figure 3.4. Estimated variation of nondimensional 95% cooling time, $3\tau\alpha/L^2$, with Bi.

Comparison of Model to Exact Results

We have now generated estimates of the plate's cooling rate for a broad range of $\bar{h}$. How good are the results? The well-known Fourier series solution for the temperature of a slab can be approximated by its first term for times larger than about $0.3\,L^2/\alpha$:

$$\frac{T - T_\infty}{T_0 - T_\infty} \cong \left(\frac{2\sin\lambda}{\lambda + \sin\lambda\cos\lambda}\cos(\lambda x/L)\right)\exp\left(-\lambda^2\frac{t\alpha}{L^2}\right) \tag{3.22}$$

where λ is the smallest positive number satisfying $\lambda\tan\lambda = Bi$. We may make several observations:

1. The cooling is exponential for all values of the Biot number for large time.
2. The time constant is $\tau = \frac{1}{\lambda^2}\frac{L^2}{\alpha}$.
3. For any position, x, within the plate, the process has the same time constant even though temperature varies in x; thus, the time constant we estimated using Eq. (3.16) is not much affected by the particular point at which the "average" temperature occurs.
4. In the limit $Bi \ll 1$, it can be shown that $\lambda^2 \cong Bi$. The cosine term is therefore approximately equal to unity for all x/L. Additional algebra then shows that this solution corresponds exactly to the lumped capacity solution Eq. (3.11). The

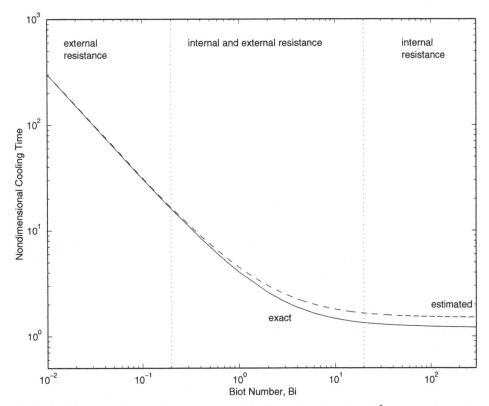

Figure 3.5. Exact variation of nondimensional 95% cooling time, $3\tau\alpha/L^2$, compared to estimated variation.

    lumped capacity solution is accurate to within about 3% for Bi < 0.1 and 12% for Bi < 0.5 [3.2, Table 3].

5. In the limit Bi ≫ 1, it can be shown that $\lambda^2 \cong \pi^2/4$. The corresponding time constant is $\tau = \frac{4}{\pi^2}\frac{L^2}{\alpha} = 0.41\frac{L^2}{\alpha}$ as opposed to the value $0.5\frac{L^2}{\alpha}$ that we estimated. In other words, our estimate is within 20% of the correct value. It should be noted, however, that if we had based the conduction resistance, Eq. (3.6), on a different conduction length, the disagreement could be significantly larger.

The exact and estimated solutions are compared in Figure 3.5. Unless the Biot number is ≪ 1, the agreement is not this good for cooling times less than $0.3\,L^2/\alpha$, which corresponds to roughly the first 50% of the cooling process.

3.2.2 More on Thermal Circuits

Resistance–capacitance models can easily be extended to situations involving more than one lumped thermal capacitance and many thermal resistances. The application of energy conservation to each capacitance in the circuit will then lead to a set of coupled first-order differential equations describing the temperature response of the system.

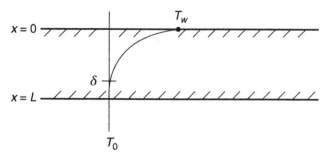

Figure 3.6. Initial temperature response of a heated slab.

Any thermal capacitance in such a circuit will have one side "grounded" at an arbitrary reference temperature. This reference temperature is the thermodynamic datum against which internal energy is calculated. For example, a lumped object with a constant specific heat capacity has $U = mc_p(T - T_{ref})$ for T_{ref} the datum. Since the internal energy appears in the energy balance equations as $dU/dt = mc_p dT/dt$, the choice of T_{ref} is not significant. For the circuit shown in Figure 3.3, we used $T_{ref} = T_\infty$, but any other value would have been acceptable.

When a thermal capacitance gains or loses heat through more than one thermal resistance, each resistance affects the time constant. When several capacitances are present at different temperatures, they are coupled by thermal resistances, and each one is separately grounded. Each capacitance affects the overall response, and the response will have the same number of time constants as there are thermal capacitances. For example, a system with two interacting capacitances will have a step-change response of the form

$$T(t) = T_{steady} + ae^{-t/\tau_1} + be^{-t/\tau_2} \tag{3.23}$$

Analytical expressions for the two time constants, τ_1 and τ_2, depend on the specific configuration and are usually relatively complicated (see, e.g., Lienhard and Lienhard [3.3, Section 5.2]). When the time constants of the two capacitances have very different magnitudes, the transient can be simplified into two distinct time periods of response. In some cases, and especially as the number of capacitances increases, the best approach may be to set up the first-order equations and to solve them numerically using standard software. Problem 3.1 explores interacting capacitances further.

3.2.3 Resistance and Capacitance for Short Time Scales: Propagation of Temperature Changes

Let us look at the initial response to heating of an object that is initially at a uniform temperature, such as a concrete floor after a large tank of hot liquid has been dumped onto it. We will assume that convection resistance is small compared to conduction resistance (Bi $\gg$ 1), so that the surface of the floor is brought abruptly to the liquid temperature and that the liquid pool is deep enough not to cool quickly. Initially, only the upper part of the concrete shows any temperature change (Fig. 3.6). Within a distance of δ below the surface, the temperature drops from the surface temperature T_w to the initial temperature, T_0. This distance grows in time as the slab continues to warm up.

Can we estimate how fast this region grows and how long until the entire thickness, L, of the slab shows a temperature change? Obviously, the well-known analytical solution for the temperature response of a semi-infinite body would apply here. However, let us try instead to model the response using an energy balance and an approximation to the temperature profile. In this case, because the temperature profile is spreading in time, it is convenient to think in terms of the total temperature difference, $(T_w - T_0)$, rather than a mean internal temperature as we did in the previous section.

The heat flow into the slab serves to raise the internal energy, U, according to the First Law:

$$\frac{d}{dt}U = Q \tag{3.24}$$

At any time, the internal energy that has been added to the slab can be found by integrating the temperature profile

$$U = \int_0^\delta \rho c_p A (T - T_0) \, dx \tag{3.25}$$

where we take the initial temperature, T_0, as the thermodynamic datum. As more of the slab warms, more of its thermal capacitance is involved in storing the heat transferred into it.

The heat flow rate into the slab can be found by differentiating the temperature profile

$$Q = -kA \left.\frac{dT}{dx}\right|_{x=0} \tag{3.26}$$

This also defines the conduction resistance, since $Q = (T_w - T_0)/R_{cond}$. Because the temperature profile is steeply curved, most of the heat entering the slab does not flow the full distance δ, and most of the thermal resistance will lie near the surface. We therefore expect that at any particular time the conduction resistance is roughly

$$R_{cond} \approx \frac{\delta/2}{kA} \tag{3.27}$$

but this can be checked after approximating the temperature profile and calculating Q.

The temperature profile may be written in terms of the overall temperature difference and a shape function $f(x)$ that decays from 1 to 0 as x increases from 0 to δ:

$$(T - T_0) = (T_w - T_0) f(x) = \Delta T \, f(x) \tag{3.28}$$

Thus,

$$U = \rho c_p A \Delta T \int_0^\delta f(x) \, dx \tag{3.29}$$

and

$$Q = -kA\Delta T f'(0) \tag{3.30}$$

The form of $f(x)$ will depend on the boundary conditions we need to satisfy. So far, two have been mentioned: $f(0) = 1$ and $f(\delta) = 0$. These set the temperature at two points. A straight line, $f(x) = 1 - x/\delta$, meets both conditions; however, it does not reach zero slope at $x = \delta$, as does the real temperature profile. This boundary condition is important because the reason for a zero slope is that all of the heat entering at $x = 0$ has been absorbed by the thermal capacitance of the region $x \le \delta$; to omit it would be to ignore the heat storage in upper part of the slab, as if it had *no* active thermal capacitance. To pick up this additional boundary condition, we can use a parabola instead, choosing coefficients that meet the two original boundary conditions and the requirement that $f'(\delta) = 0$:

$$f(x) = 1 - 2\left(\frac{x}{\delta}\right) + \left(\frac{x}{\delta}\right)^2 \tag{3.31}$$

Upon substituting this function into Eq. (3.29) we get

$$U \approx \rho c_p A\Delta T \frac{\delta}{3} \tag{3.32}$$

showing the increase in the slab's internal energy as δ grows. Further, we find from Eq. (3.30) that

$$Q \approx \frac{2kA\Delta T}{\delta} = \frac{\Delta T}{(\delta/2)/kA} = \frac{\Delta T}{R_{\text{cond}}} \tag{3.33}$$

which shows R_{cond} to be consistent with our estimate, Eq. (3.27).

Substitution of our expressions for U and Q into Eq. (3.24) gives

$$\delta \frac{d\delta}{dt} = \frac{d}{dt}\left(\rho c_p A\Delta T \frac{\delta}{3}\right) = \frac{2kA\,\Delta T}{\delta} \tag{3.34}$$

or

$$\frac{d}{dt}\left(\frac{\delta^2}{2}\right) = \frac{6k}{\rho c_p} = 6\alpha \tag{3.35}$$

On integration, we have

$$\delta = \sqrt{12\alpha t} = 3.46\sqrt{\alpha t} \tag{3.36}$$

The penetration of the temperature disturbance is proportional to $\sqrt{\alpha t}$. In physical terms, thermal resistance increases as the disturbance travels further in, slowing the rate at which the internal energy of the slab increases.

The heat flux into the slab may be found by substituting Eq. (3.36) into Eq. (3.33):

$$q_w = \frac{Q}{A} = \frac{k\Delta T}{\sqrt{3\alpha t}} \tag{3.37}$$

The exact solution for q_w has a π in place of the 3, so our model appears to work quite well. Note that the heat flux decreases steadily in time because the thermal

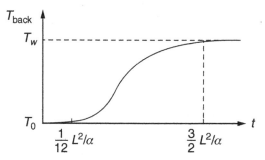

Figure 3.7. Temperature response at the back of an adiabatic heated slab.

resistance rises as δ grows. From the equation for q_w, we have:

$$R_{cond} = \frac{\sqrt{3\alpha t}}{kA} \tag{3.38}$$

Note that this resistance is *time-dependent*.

We can estimate the time required for the temperature disturbance to propagate to the back of the slab by setting $\delta \approx L$:

$$L = \sqrt{12\alpha t_L} \tag{3.39}$$

or

$$t_L = \frac{1}{12}\frac{L^2}{\alpha} \tag{3.40}$$

Thus, the back of the slab stays at T_0 until $t \approx L^2/(12\alpha)$, after which time its temperature rises. If the back of the slab is adiabatic, it will eventually warm to T_w. From the discussion earlier in this section, we know that steady state will be reached in about three time constants, or $t_{ss} \approx 3L^2/(2\alpha)$. The back temperature approaches T_w as a decaying exponential function for times greater than about $0.2L^2/\alpha$ (Fig. 3.7).

3.2.4 The Fourier Number

All of our estimates of cooling time and propagation time have turned out to be proportional to L^2/α. The quantity $t_d = L^2/\alpha$ is the characteristic time for heat to diffuse a distance L. It can be used to form a dimensionless time, called the *Fourier number*,

$$\text{Fo} \equiv \frac{t}{t_d} = \frac{\alpha t}{L^2} \tag{3.41}$$

which compares the current time to the time required for heat to diffuse over the characteristic length. When Fo is small, say less than 0.05 or so, the temperature disturbance has traveled less than the distance L and is confined to the surface of the body. The body is effectively semi-infinite. When Fo is large compared to 1, the body has reached steady state. For times in between, the temperature distribution may be found using Fourier series solutions — and for Fo > 0.2 or so, only the first term of the series is significant, as discussed in the next section.

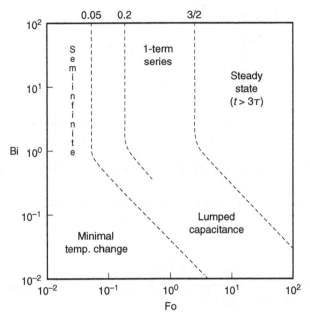

Figure 3.8. Type of conduction solution appropriate for various values of Fo and Bi in a slab.

Figure 3.8 shows the nature of the conduction solution for a slab in various ranges of Fourier number and Biot number.

3.3 Use of Series Solutions for Unsteady Conduction

Unsteady heat conduction in solids of uniform thermal conductivity is described by the following form of the energy equation, which is also referred to as the heat equation,

$$\rho c_p \underbrace{\frac{\partial T}{\partial t}}_{\text{storage}} = \underbrace{k\nabla^2 T}_{\text{conduction}} + \underbrace{\dot{q}_v}_{\text{generation}} \tag{3.42}$$

in which $\dot{q}_v$ is volumetric heat generation (W/m^3). In this equation, heat capacitance is represented by the first term and thermal resistance is represented by the second term; both are assumed to be distributed throughout the solid.

The heat equation can be integrated for a wide variety of situations, and during the past two centuries an enormous number of cases have been treated analytically. Several dozen of these solutions are of significant general value, and several thousand more are of consistent use in specialized situations. If the geometry and boundary conditions of a problem are of a regular form, one should always assume that an analytical solution has already been found by someone else. A number of books have cataloged these solutions, although none is more consistently useful than that by Carslaw and Jaeger [3.4].

When finite dimensional bodies are involved, analytical solutions of the unsteady heat equation often take the form of an infinite series, if the geometry and

Table 3.1. *Forms of one-dimensional series solutions.*

	A_n	f_n	D_n
Slab	$\dfrac{2\sin\lambda_n}{\lambda_n + \sin\lambda_n\cos\lambda_n}$	$\cos\left(\lambda_n\dfrac{x}{L}\right)$	$A_n\cdot\dfrac{\sin\lambda_n}{\lambda_n}$
Cylinder	$\dfrac{2J_1(\lambda_n)}{\lambda_n\left[J_0^2(\lambda_n) + J_1^2(\lambda_n)\right]}$	$J_0\left(\lambda_n\dfrac{r}{R}\right)$	$2A_n\cdot\dfrac{J_1(\lambda_n)}{\lambda_n}$
Sphere	$2\dfrac{\sin\lambda_n - \lambda_n\cos\lambda_n}{\lambda_n - \sin\lambda_n\cos\lambda_n}$	$\left(\dfrac{R}{\lambda_n r}\right)\sin\left(\dfrac{\lambda_n r}{R}\right)$	$3A_n\cdot\dfrac{\sin\lambda_n - \lambda_n\cos\lambda_n}{\lambda_n^3}$

boundary conditions allow separation of variables. These series are most useful when the Fourier number is greater than 0.05 or so; for smaller values, the temperature changes tend to be confined to the region near the surface of the body, and semi-infinite body solutions are usually more convenient to use, as discussed in Section 3.4.

For cases of one-dimensional transient convective cooling or heating (with $\dot{q}_v = 0$), the series solutions have the form

$$\Theta = \frac{T - T_\infty}{T_0 - T_\infty} = \sum_{n=1}^{\infty} A_n \exp\left(-\lambda_n^2 \text{Fo}\right) f_n \tag{3.43}$$

where f_n is a function of the relevant coordinate and λ_n is an eigenvalue that depends on the Biot number. This equation would apply, for example, to a sphere of radius R initially at a uniform temperature T_0 that is cooled from its surface by convection to a medium at constant temperature T_∞. The eigenvalues increase with increasing n, so that only the first term of the series is nonnegligible when the Fourier number is greater than about 0.2 to 0.4.

Table 3.1 gives the functions in the series for the case of a slab of thickness $2L$ cooled by convection at $x = \pm L$, a long cylinder of radius R, and a sphere of radius R as a function of the Biot number. For the slab, the Biot number is based on the half-thickness L, and for the cylinder and the sphere, it is based on R. The same dimensions are used in the Fourier number in the series; for example, the Fourier number of the slab is defined as $\alpha t/L^2$. In general, the eigenvalue λ_n is the nth root of a transcendental equation; in the case of the sphere, that equation is

$$\lambda \cot\lambda = 1 - \text{Bi}_R \equiv 1 - \frac{\bar{h}R}{k} \qquad \text{for a sphere} \tag{3.44}$$

where $\bar{h}$ is the heat transfer coefficient at the surface of the sphere and k is the thermal conductivity of the sphere.[5]

The direct use of these series solutions in problem solving is discussed at length in most undergraduate textbooks. As an aid in manipulating the series, it is common to plot the value of Θ as a function of the Fourier number for various values of the Biot number at a given position (x/L or r/R). Such temperature response charts were

---

[5] For a slab, $\text{Bi}_L = \lambda \tan\lambda$; for a cylinder, $\lambda J_1(\lambda) = \text{Bi}_R J_0(\lambda)$ where J_0 and J_1 are Bessel functions of the first kind.

widely used before the 1970s, and some very extensive collections of these charts can be found in the older literature. Schneider, for example, published more than 120 of them in 1963 [3.5]. These charts go well beyond what is normally found in undergraduate textbooks, including such items as multilayered configurations and sinusoidal surface heat fluxes.

The main value of such temperature response charts is at intermediate Fourier numbers where the body is not semi-infinite, but for which several terms of the series must be summed to obtain an accurate solution. Roughly speaking, this corresponds to $0.05 \lesssim \mathrm{Fo} \lesssim 0.3$; the precise limits, of course, depend upon the allowable error and the Biot number.

One-dimensional transient solutions can be extended to a variety of other cases through the use of superposition to accommodate other boundary conditions and the presence of volumetric heating. Simple multidimensional cases can sometimes be represented as products of one-dimensional solutions. In other cases, analysis will lead to multiple series solutions, with a summation for each coordinate, which can rapidly become inconvenient to use. All of these topics are treated in standard texts on heat conduction, and they may be of use in constructing more sophisticated models of physical problems.

3.3.1 Mean Temperature and Total Heat Transfer

When an object is cooled or heated, it is often useful to know the total heat transferred up to some given time, or, equivalently, to know the mean temperature of the object at that time.

Thermodynamic considerations show that the maximum heat removal possible for an object of volume $\mathcal{V}$ cooling from T_0 to T_∞ and having uniform ρc_p is

$$\int_0^\infty Q(t)\, dt = \rho c_p \mathcal{V}\, (T_0 - T_\infty) \tag{3.45}$$

The total heat loss that has occurred by a time t is

$$\int_0^t Q(t)\, dt = \rho c_p \mathcal{V}\, [T_0 - \overline{T}(t)] \tag{3.46}$$

for $\overline{T}(t)$ the spatially averaged temperature of the body. Hence, we may write

$$\overline{\Theta} \equiv \frac{\overline{T} - T_\infty}{T_0 - T_\infty} = 1 - \frac{\displaystyle\int_0^t Q(t)\, dt}{\displaystyle\int_0^\infty Q(t)\, dt} \equiv 1 - \Phi \tag{3.47}$$

Both $\overline{\Theta}$ and Φ are functions of Bi and Fo. By spatially averaging the series solution Eq. (3.43) to obtain $\overline{T}$ (or by differentiating to get Q and then integrating), one finds that

$$\overline{\Theta} = 1 - \Phi = \sum_{n=1}^{\infty} D_n \exp\left(-\lambda_n^2 \mathrm{Fo}\right) \tag{3.48}$$

Expressions for D_n are given in Table 3.1 for one-dimensional convective cooling or heating of a slab, a cylinder, and a sphere. Response charts for $\Phi(\text{Bi}, \text{Fo})$ are also common in the literature.

3.3.2 One-Term Solutions

For relatively large values of the Fourier number, greater than 0.2 to 0.4, only the first term in the series makes a significant contribution to the sum, and it is possible to approximate the temperature distribution by the first term alone:

$$\Theta \approx A_1 \cdot f_1 \cdot \exp\left(-\lambda_1^2 \text{Fo}\right) \tag{3.49}$$

The mean temperature can also be approximated with a single term:

$$\overline{\Theta} = 1 - \Phi \approx D_1 \exp\left(-\lambda_1^2 \text{Fo}\right) \tag{3.50}$$

Table 3.2 lists the values of λ_1, A_1, and D_1 for slabs, cylinders, and spheres as a function of the Biot number. The one-term solution's error in Θ is less than 0.1% for a sphere with Fo ≥ 0.28 and for a slab with Fo ≥ 0.43. The errors are largest for Biot numbers near unity.

Ostrogorsky [3.2] has provided approximate equations for A_1, λ_1, and D_1 as a function of Biot number (Table 3.3). These can be used for all Biot number in the range Fo > 0.2 with an error less than 1%.

3.3.3 Modeling Tactics

Most undergraduate textbooks include useful examples of adapting the one-dimensional series solutions as approximations to more complex situations. Some basic tactics include the following:

- Approximating nearly cylindrical, spherical, or planar objects as exact cylinders, spheres, or planes. For example, an egg-shaped object might be modeled as a sphere. If the minor diameter (smallest diameter) were used in estimating a cooling time, both the thermal capacitance and internal thermal resistance are taken to be less than the actual case so that the cooling time calculated would be a lower bound on the actual cooling time. An upper bound would be obtained by using the major (maximum) diameter. It should be clear that some parts of an irregularly shaped object will cool more rapidly than others.
- A thermal resistance external to an object being cooled may be treated using convective solutions if it has no associated thermal capacitance. The apparent convection coefficient of the resistance is simply the reciprocal of the thermal resistance per unit area surface area of the cooled object. External resistance in series or parallel may be replaced by equivalent total resistance before finding the apparent convection coefficient.
- Upper and lower bounds on surface temperature may be used to bound the time required for temperature change. For example, the surface temperature

Table 3.2. *One-term coefficients for one-dimensional convective heat transfer.*

Bi	Plate			Cylinder			Sphere		
	λ_1	A_1	D_1	λ_1	A_1	D_1	λ_1	A_1	D_1
0.01	0.09983	1.0017	1.0000	0.14124	1.0025	1.0000	0.17303	1.0030	1.0000
0.05	0.22176	1.0082	0.9999	0.31426	1.0124	0.9999	0.38537	1.0150	1.0000
0.10	0.31105	1.0161	0.9998	0.44168	1.0246	0.9998	0.54228	1.0298	0.9998
0.15	0.37788	1.0237	0.9995	0.53761	1.0365	0.9995	0.66086	1.0445	0.9996
0.20	0.43284	1.0311	0.9992	0.61697	1.0483	0.9992	0.75931	1.0592	0.9993
0.25	0.48009	1.0382	0.9988	0.68559	1.0598	0.9988	0.84473	1.0737	0.9990
0.30	0.52179	1.0450	0.9983	0.74646	1.0712	0.9983	0.92079	1.0880	0.9985
0.40	0.59324	1.0580	0.9971	0.85158	1.0931	0.9970	1.05279	1.1164	0.9974
0.50	0.65327	1.0701	0.9956	0.94077	1.1143	0.9954	1.16556	1.1441	0.9960
0.60	0.70507	1.0814	0.9940	1.01844	1.1345	0.9936	1.26440	1.1713	0.9944
0.70	0.75056	1.0918	0.9922	1.08725	1.1539	0.9916	1.35252	1.1978	0.9925
0.80	0.79103	1.1016	0.9903	1.14897	1.1724	0.9893	1.43203	1.2236	0.9904
0.90	0.82740	1.1107	0.9882	1.20484	1.1902	0.9869	1.50442	1.2488	0.9880
1.00	0.86033	1.1191	0.9861	1.25578	1.2071	0.9843	1.57080	1.2732	0.9855
1.10	0.89035	1.1270	0.9839	1.30251	1.2232	0.9815	1.63199	1.2970	0.9828
1.20	0.91785	1.1344	0.9817	1.34558	1.2387	0.9787	1.68868	1.3201	0.9800
1.30	0.94316	1.1412	0.9794	1.38543	1.2533	0.9757	1.74140	1.3424	0.9770
1.40	0.96655	1.1477	0.9771	1.42246	1.2673	0.9727	1.79058	1.3640	0.9739
1.50	0.98824	1.1537	0.9748	1.45695	1.2807	0.9696	1.83660	1.3850	0.9707
1.60	1.00842	1.1593	0.9726	1.48917	1.2934	0.9665	1.87976	1.4052	0.9674
1.70	1.02725	1.1645	0.9703	1.51936	1.3055	0.9633	1.92035	1.4247	0.9640
1.80	1.04486	1.1695	0.9680	1.54769	1.3170	0.9601	1.95857	1.4436	0.9605
1.90	1.06136	1.1741	0.9658	1.57434	1.3279	0.9569	1.99465	1.4618	0.9570
2.00	1.07687	1.1785	0.9635	1.59945	1.3384	0.9537	2.02876	1.4793	0.9534
2.20	1.10524	1.1864	0.9592	1.64557	1.3578	0.9472	2.09166	1.5125	0.9462
2.40	1.13056	1.1934	0.9549	1.68691	1.3754	0.9408	2.14834	1.5433	0.9389
2.60	1.15330	1.1997	0.9509	1.72418	1.3914	0.9345	2.19967	1.5718	0.9316
2.80	1.17383	1.2052	0.9469	1.75794	1.4059	0.9284	2.24633	1.5982	0.9243
3.00	1.19246	1.2102	0.9431	1.78866	1.4191	0.9224	2.28893	1.6227	0.9171
3.50	1.23227	1.2206	0.9343	1.85449	1.4473	0.9081	2.38064	1.6761	0.8995
4.00	1.26459	1.2287	0.9264	1.90808	1.4698	0.8950	2.45564	1.7202	0.8830
4.50	1.29134	1.2351	0.9193	1.95248	1.4880	0.8830	2.51795	1.7567	0.8675
5.00	1.31384	1.2402	0.9130	1.98981	1.5029	0.8721	2.57043	1.7870	0.8533
6.00	1.34955	1.2479	0.9021	2.04901	1.5253	0.8532	2.65366	1.8338	0.8281
7.00	1.37662	1.2532	0.8932	2.09373	1.5411	0.8375	2.71646	1.8673	0.8069
8.00	1.39782	1.2570	0.8858	2.12864	1.5526	0.8244	2.76536	1.8920	0.7889
9.00	1.41487	1.2598	0.8796	2.15661	1.5611	0.8133	2.80443	1.9106	0.7737
10.00	1.42887	1.2620	0.8743	2.17950	1.5677	0.8039	2.83630	1.9249	0.7607
12.00	1.45050	1.2650	0.8658	2.21468	1.5769	0.7887	2.88509	1.9450	0.7397
14.00	1.46643	1.2669	0.8592	2.24044	1.5828	0.7770	2.92060	1.9581	0.7236
20.00	1.49613	1.2699	0.8464	2.28805	1.5919	0.7542	2.98572	1.9781	0.6922
25.00	1.51045	1.2710	0.8400	2.31080	1.5954	0.7427	3.01656	1.9856	0.6766
30.00	1.52017	1.2717	0.8355	2.32614	1.5973	0.7348	3.03724	1.9898	0.6658
50.00	1.54001	1.2727	0.8260	2.35724	1.6002	0.7183	3.07884	1.9962	0.6434
100.00	1.55525	1.2731	0.8185	2.38090	1.6015	0.7052	3.11019	1.9990	0.6259
∞	1.57080	1.2732	0.8106	2.40483	1.6020	0.6917	3.14159	2.0000	0.6079

Data follow H.D. Baehr and K. Stephan [3.6]

Table 3.3. *Approximate equations for parameters of the one-term solutions [3.2].*

	A_1	λ_1	D_1
Slab	$1 + \dfrac{0.273}{\left(1 + 2.42/Bi^{1.5}\right)^{2/3}}$	$\dfrac{\pi/2}{\left(1 + 2.62/Bi^{1.07}\right)^{0.468}}$	$1 - \dfrac{0.189}{\left(1 + 3.8/Bi^{1.116}\right)^{1.62}}$
Cylinder	$1 + \dfrac{0.602}{\left(1 + 4.8/Bi^{1.08}\right)^{0.61}}$	$\dfrac{2.4048}{\left(1 + 3.28/Bi^{1.125}\right)^{0.446}}$	$1 - \dfrac{0.308}{\left(1 + 2.58/Bi^{1.08}\right)^{2.35}}$
Sphere	$1 + \dfrac{1}{\left(1 + 8.87/Bi^{1.76}\right)^{0.568}}$	$\dfrac{\pi}{\left(1 + 4.1/Bi^{1.18}\right)^{0.4238}}$	$1 - \dfrac{0.392}{\left(1 + 2.36/Bi^{1.09}\right)^{2.85}}$

of a cooled object may rise during the cooling process. A lower bound on the cooling time would be obtained by fixing the surface temperature at its initial value. If the final surface temperature is known, an upper bound on cooling time is obtained by fixing the surface temperature at the final value throughout the cooling process.

- A slab boundary that is adiabatic may be regarded as a line of symmetry for an effective slab which is twice as thick and being cooled equally from both sides. Further, if one boundary of a slab has a far lower rate of heat transfer than the other side (e.g., it is heavily insulated or inefficiently cooled), that boundary may be approximated as adiabatic during early stages of the process; however, the adiabatic approximation may or may not become inappropriate as steady state is approached.

Example 3.3 *Cooling glass globes* To form thin glass globes, hot glass is blown into copper molds. The glass is initially at 1800°F (982°C) and the mold is at an essentially uniform temperature of 500°F (260°C). The glass wall thickness is $t_g = 0.2$ inch (5.1 mm) and the glass is in good contact with the copper. The copper mold thickness is $t_{Cu} = 0.5$ inch (12.7 mm). The exterior of the copper mold is exposed to still air near room temperature. We would like to obtain upper and lower bounds on the time required for the temperature of the inner surface of the glass to reach 1000°F (538°C).

The size of the globes is not stated, but we may assume their radius, R_{globe}, to be large compared to the thickness of the glass, t_g, so that glass curvature can be neglected. We will also neglect contact resistance between the glass and the mold and neglect radiation in the glass. For the glass, $k = 0.9$ W/m·K, $\rho = 2500$ kg/m³, and $c_p = 880$ J/kg·K; for the copper, $k = 380$ W/m·K, $\rho = 8900$ kg/m³, and $c_p = 380$ J/kg·K.

To understand what governs the cooling process, we may start by identifying the important thermal capacitances and resistances. The heat capacitance of the glass drives the heat transfer problem, and the major heat transfer will be from the glass into the copper mold. What about the capacitance of air inside the globe? We can compare the thermal capacitance of the air to that of the glass by considering the ratio of the two:

$$\frac{(\rho c_p)_{\text{air}}\left(\frac{4}{3}\pi R_{\text{globe}}^3\right)}{(\rho c_p)_{\text{glass}}\left(4\pi R_{\text{globe}}^2 t_g\right)} = \left(\frac{1.2 \times 10^3 \text{ J/m}^3 \cdot \text{K}}{2.2 \times 10^6 \text{ J/m}^3 \cdot \text{K}}\right)\frac{R_{\text{globe}}}{3t_g} = 1.8 \times 10^{-4}\frac{R_{\text{globe}}}{t_g} \quad (3.51)$$

Unless the globes are very large (more than several meters in diameter), the air capacitance will be negligible. This means that we can ignore convective heat transfer to the air inside. Further, the inside surface of the glass faces glass at the same temperature, so no net radiation will occur. The inside surface of the glass may therefore be approximated as adiabatic.

Next we may consider the thermal resistances of the glass, the mold, and the convection/radiation cooling to the environment outside the mold. We can begin by considering times long enough that temperature changes have propagated all the way through the glass and the mold (e.g., $\delta > t_g$ as in Section 3.2.3):

$$R_{\text{Cu}} \approx \frac{t_{\text{Cu}}}{k_{\text{Cu}}A} \quad (3.52\text{a})$$

$$R_{\text{glass}} \approx \frac{t_g}{2k_g A} \quad (3.52\text{b})$$

$$R_{\text{outside}} = \frac{1}{(h_{\text{nc}} + h_r)A} \quad (3.52\text{c})$$

where A is the surface area of the mold, and the conductance of natural convection and radiation, $(h_{\text{nc}} + h_r)$, will have typical values of 5 to 20 W/m^{2}K. A look at the numbers quickly shows that the resistances are widely spread:

$$R_{\text{Cu}} \ll R_{\text{glass}} \ll R_{\text{outside}} \quad (3.53)$$

This suggests that we can model the problem by neglecting the temperature gradient in the copper mold as it warms up, so that the only copper temperature is a function of time. The glass will have significant temperature gradients as it cools. The resistance outside the mold is very large, so that the copper mold may warm up considerably relative to the surroundings before it eventually cools off.

We need to think about the difference in the time scale for the glass to transfer heat to the mold and for the mold and glass to finally cool down. The final cooling can be estimated using a lumped capacitance model, since the outside convection/radiation resistance is much larger than the internal resistance of the glass and mold. Then, with Eq. (3.12),

$$\tau = [(mc_p)_{\text{Cu}} + (mc_p)_{\text{glass}}]R_{\text{outside}} \quad (3.54)$$

$$= \{[(\rho c_p)t]_{\text{Cu}} + [(\rho c_p)t]_{\text{glass}}\}\frac{1}{(h_{\text{nc}} + h_r)} \quad (3.55)$$

$$\approx \{(3.4 \times 10^6)(0.0127) + (2.2 \times 10^6)(0.0051)\}\frac{1}{20} \quad (3.56)$$

$$= 2720 \text{ s} \quad (3.57)$$

This is roughly 45 minutes, and for times that are small compared to τ, the mold will have given up little heat to the surroundings. The time scale on which the glass cools toward the mold temperature can be estimated by finding the time at which the Fourier number reaches 3/2 (Section 3.2.4):

$$t = \text{Fo}\,\frac{t_g^2}{\alpha_g} = \frac{3}{2}\frac{(0.0051)^2}{(4.1 \times 10^{-7})} = 95 \text{ sec} \tag{3.58}$$

Thus, the thermal interaction between the glass and the mold occurs much faster than the cooling into the environment. We may model the outside surface of the mold as being adiabatic during the time period of interest.

The remaining problem is to understand how the mold temperature varies while the glass cools. On initial contact, the interface between the glass and the copper immediately comes to a temperature of 549°F (287°C), as will be shown in Example 3.4, and the mold temperature will start to rise further after less than 1 second. The mold temperature will rise only until the glass and the copper reach thermal equilibrium, at which they will have the same final temperature. Neglecting heat transfer to the surroundings, this final temperature, T_f, may be determined from the first law of thermodynamics:

$$[(\rho c_p)t]_{\text{Cu}} A (T_f - 260) = [(\rho c_p)t]_{\text{glass}} A (982 - T_f) \tag{3.59}$$

which gives $T_f = 399°C$. Hence, the temperature of the copper mold surface touching the glass varies between a lower bound of 287°C and an upper bound of 399°C.

We are now able to make upper and lower bounds on the time needed for the inside of the glass to cool to 538°C: we can calculate a lower bound on cooling time using the lower bound on the copper temperature and the upper bound on time using the upper bound on temperature. Because the adiabatic inside surface of the glass has had a significant temperature change, we expect that Fo ≥ 0.2. Therefore, we use the one-term approximation, treating the adiabatic surface as the centerline of a slab $(x/L = 0)$ and using the solution for a fixed surface temperature (Bi $\longrightarrow \infty$). From Eq. (3.49) and Table 3.2:

$$\Theta = \frac{538 - T_{\text{Cu}}}{982 - T_{\text{Cu}}} \approx 1.2732 \cdot \exp[-(1.5708)^2 \, \text{Fo}]. \tag{3.60}$$

The lower bound, $T_{\text{Cu}} = 287°C$, gives $\Theta = 0.361$, Fo $= 0.511$, and $t_{\text{lb}} = 32.5$ s. The upper bound, $T_{\text{Cu}} = 399°C$, gives $\Theta = 0.238$, Fo $= 0.679$, and $t_{\text{ub}} = 43.2$ s.

These bounds are quite close, and we might summarize by saying that the inside temperature will reach 538°C after 38 ± 5 seconds.

3.4 Using Semi-infinite Body Solutions

A body may be regarded as remaining semi-infinite while a temperature disturbance applied to one surface does not extend to other surfaces of the body and is not significantly affected by curvature of the surface. In such cases, the physical dimensions of the body do not provide an appropriate characteristic length scale for heat conduction. Temperature disturbances will initially penetrate as $\sqrt{\alpha t}$ in most cases. In a few particular situations, analytical solutions allow the penetration depth to be calculated precisely so as to test the applicability of semi-infinite body results. For example, a slab with a step-changed temperature on one boundary will be semi-infinite for times short enough that $\delta = 3.65\sqrt{\alpha t}$ is less than the slab thickness.

Most simple semi-infinite body solutions of the heat equation are for plane slabs. A few such results are summarized in Table 3.4.[6] The table includes the temperature response of semi-infinite bodies ($x \geq 0$) experiencing several types of temperature variations at boundary $x = 0$. Also included are equations for the heat flux at $x = 0$ and the 99% penetration depth, which is the distance from $x = 0$ at which the temperature change from the initial temperature is 1% of the temperature difference between the surface and the initial temperature. A physical model for the penetration depth was discussed in Section 3.2.3.

Note that the property group $\sqrt{k\rho c_p}$ appears in the equations for time-dependent heat flux. Some authors refer to $k\rho c_p$ as the *heat diffusivity* (not to be confused with the thermal diffusivity, α); however, this designation is uncommon. (Further, $\sqrt{k\rho c_p}$ is sometimes called *thermal effusivity*.) As discussed at the beginning of this chapter, the value of ρc_p does not vary much among nonporous solids. Consequently, the value of $\sqrt{k\rho c_p}$ depends mainly on a material's thermal conductivity.

Step change in wall temperature. This is the canonical semi-infinite body solution. The solution appears in Table 3.4 in terms of the complementary error function, erfc(x), which can be defined in terms of the error function, erf(x), as erfc(x) = $1 - \mathrm{erf}(x)$. The error function is defined as

$$\mathrm{erf}(x) \equiv \frac{2}{\sqrt{\pi}} \int_0^x e^{-x^2}\, dx. \tag{3.61}$$

The derivative of erfc(x) is

$$\frac{d}{dx}\,\mathrm{erfc}(x) = -\frac{2}{\sqrt{\pi}} e^{-x^2}. \tag{3.62}$$

An asymptotic expansion is available for erfc(x) at large arguments

$$\mathrm{erfc}(x) \sim \frac{1}{\sqrt{\pi}} e^{-x^2} \left(\frac{1}{x} - \frac{1}{2x^3} + + \frac{3}{4x^5} - \cdots \right), \quad x \longrightarrow \infty \tag{3.63}$$

Step change in wall heat flux. This solution is almost as well known as the preceding one. In this case, however, the wall temperature rises steadily while heating

---

[6] These solutions and a number of others are derived in Carslaw and Jaeger [3.4, Chapter 2]. A result known as Duhamel's theorem gives solutions for arbitrary time variations of the boundary temperature in terms of an integral.

Table 3.4. *Some useful planar semi-infinite body solutions ($x \geq 0$, $t > 0$). Initial body temperature of T_0.*

Case	Temperature response	Penetration depth, heat flux
Step change in wall temperature from T_0 to T_w at $t = 0$	$\dfrac{T(x,t) - T_0}{T_w - T_0} = \text{erfc}\left(\dfrac{x}{2\sqrt{\alpha t}}\right)$	$\delta(t) = 3.65\sqrt{\alpha t}$ $q_w(t) = \sqrt{\dfrac{k\rho c_p}{\pi t}}(T_w - T_0)$
Step change in wall heat flux from 0 to q_w at $t = 0$	$T(x,t) - T_0 = \dfrac{q_w}{k}\left[2\sqrt{\dfrac{\alpha t}{\pi}}\exp\left(-\dfrac{x^2}{4\alpha t}\right) - x\,\text{erfc}\left(\dfrac{x}{2\sqrt{\alpha t}}\right)\right]$	$\delta(t) = 3.20\sqrt{\alpha t}$ $q_w = \text{constant}$
Step change in environmental temperature from T_0 to T_∞ at $t = 0$ with convection between environment and body surface	$\dfrac{T(x,t) - T_\infty}{T_0 - T_\infty} = \text{erf}\left(\dfrac{\zeta}{2}\right) + \exp(\beta\zeta + \beta^2)\,\text{erfc}\left(\dfrac{\zeta}{2} + \beta\right)$ $\zeta = x/\sqrt{\alpha t} \qquad \beta = \bar{h}\sqrt{\alpha t}/k$	$q_w(t) = \bar{h}(T_\infty - T_0)\,e^{\beta^2}\,\text{erfc}\,\beta$
Surface temperature varies periodically as $T_w(t) = T_0 + \Delta T\sin\omega t$ for $t > -\infty$	$T(x,t) - T_0 = \Delta T\exp\left(-x\sqrt{\dfrac{\omega}{2\alpha}}\right)\sin\left[\omega t - x\sqrt{\dfrac{\omega}{2\alpha}}\right]$	$\delta_{\text{amplitude}} = 4.61\sqrt{2\alpha/\omega}$ $q_w(t) = \sqrt{\omega(k\rho c_p)}\,\Delta T\sin(\omega t + \pi/4)$
Interfacial temperature, T_i, of two semi-infinite bodies initially at T_1 and T_2 brought into contact at $t = 0$	$\dfrac{T_i - T_1}{T_2 - T_1} = \dfrac{\sqrt{(k\rho c_p)_2}}{\sqrt{(k\rho c_p)_1} + \sqrt{(k\rho c_p)_2}}$ $\dfrac{T_j(x,t) - T_j}{T_i - T_j} = \text{erfc}\left(\dfrac{|x|}{2\sqrt{\alpha t}}\right)$ for $j = 1, 2$	$\delta_j(t) = 3.65\sqrt{\alpha_j t}$ $q_j(t) = \sqrt{\dfrac{(k\rho c_p)_j}{\pi t}}(T_i - T_j)$

progresses:

$$T_w(t) = T_0 + \frac{2q_w}{k}\sqrt{\frac{\alpha t}{\pi}} \tag{3.64}$$

The 99% penetration depth, shown in Table 3.4, is about 12% shorter than for the step-changed temperature.

Step change in environmental temperature with a convection resistance. When a semi-infinite body interacts with its environment through a convection heat transfer coefficient, the thermal resistance of the boundary layer, $1/\bar{h}$, acts as an additional parameter. The convection resistance must be compared to the conduction resistance within the body, which grows in time as the temperature change propagates farther into the body. At any time, the conduction resistance is proportional to δ/k, or $\sqrt{\alpha t}/k$ [see Eq. (3.38)]. Thus, the solution contains a time-dependent Biot number, β, defined as

$$\beta \equiv \frac{\bar{h}\sqrt{\alpha t}}{k} = \frac{\bar{h}\sqrt{t}}{\sqrt{k\rho c_p}} \approx \frac{\text{conduction resistance}}{\text{convection resistance}} \tag{3.65}$$

For sufficiently large values of β, the convection resistance is negligible, and the solution is equivalent to a semi-infinite body with a step-changed wall temperature. The temperature response is shown graphically in Figure 3.9, in terms of β and $\zeta = x/\sqrt{\alpha t}$.

The penetration depth in this case depends on β and a simple expression is not available. Clearly, $\delta < 3.65\sqrt{\alpha t}$ (the step-changed temperature result) because the convection process adds an external thermal resistance that reduces the rate of heat flow into the slab. The step-changed temperature δ provides an upper bound on the penetration depth, and it can be used to estimate a lower bound on the time during which the slab is semi-infinite. This bound is least accurate at small values of β but becomes exact as $\beta \to \infty$.

Periodically varying wall temperature. For this fourth case in Table 3.4, the wall temperature undergoes a sinusoidal time variation for $t > -\infty$. If the time variation starts at some later time, say $t = 0$, then an additional transient term must be added to the steady period response given in the table; but the transient dies out as time increases (see [3.4, Section 2.6]). The equation in Table 3.4 shows that the response is composed of a sinusoidal time variation whose amplitude decays exponentially with increasing distance into the semi-infinite body. The amplitude of variation decays to 1% of ΔT at a distance

$$\delta_{\text{amplitude}} = 4.61\sqrt{2\alpha/\omega} \tag{3.66}$$

These results can be useful in analyzing temperature changes caused by seasonal variations in the air temperature. For example, glaciers are deep masses of ice, often hundreds of meters thick. Their surface is exposed to the periodic annual variation of the weather. Given that the thermal diffusivity of ice is $\alpha_{\text{ice}} = 1.3 \times 10^{-6}$ m^2/s, and that the frequency of annual variation is $\omega = 2.0 \times 10^{-7}$ Hz, we find that

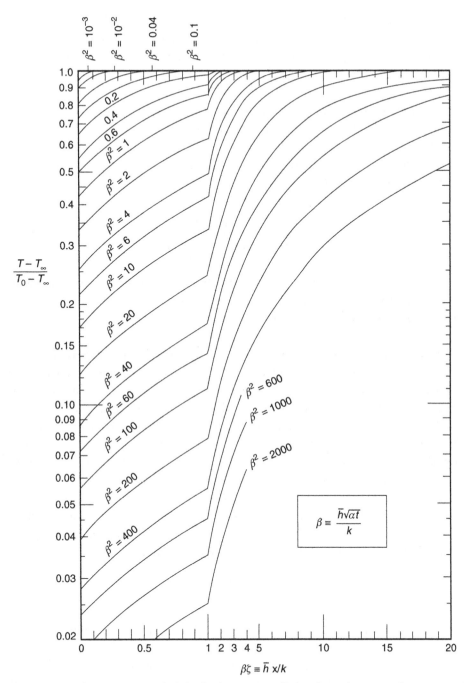

Figure 3.9. Response of a semi-infinite body to a step-change in environmental temperature with a convection boundary condition (from Lienhard & Lienhard [3.3]). Note change of scale at $\beta\zeta = 1$.

$\delta_{amplitude} = 16.7$ m, which is consistent with field observations. Beyond this depth, glacial ice stays at the annual mean temperature, T_0. For so-called polar glaciers, T_0 is below 0°C. For temperate glaciers, $T_0 > 0$°C [3.7, pp. 39–41]. What do you suppose is the condition of the interior of a temperate glacier?

Two isothermal semi-infinite bodies brought into contact. This solution can be applied to find the initial temperature response of two bodies at different temperatures, T_1 and T_2, that are brought into contact. In fact, each body will behave as a semi-infinite body which experiences a step change in its surface temperature to the interfacial temperature, T_i, given by

$$\frac{T_i - T_1}{T_2 - T_1} = \frac{\sqrt{(k\rho c_p)_2}}{\sqrt{(k\rho c_p)_1} + \sqrt{(k\rho c_p)_2}} \tag{3.67}$$

The heat fluxes and penetration depths can be computed simply using T_i for the temperature to which either body's surface is stepped. These results apply while the δ's for *both* bodies remain less than their thickness.

Spherical and Cylindrical Problems. In some cases, nonplane bodies may be approximated as planar. For example, a cylinder of radius R that is heated at its surface may be modeled as a planar semi-infinite body while $\delta \ll R$. In other situations, it may be possible to use a change of variables to transform a spherical semi-infinite problem into a planar one. Specifically, for purely radial flow in spherical coordinates (T a function of r only) if the temperature, T, is replaced by $u = rT$, the planar equation of conduction is obtained (see [3.4, Section 9.1, Section 9.10]). Some caution will be needed in transforming the boundary conditions, however.

For example, consider a volume bounded internally by a spherical surface of radius R and having an initial temperature of T_0. If the temperature of the surface $r = R$ is stepped to T_w at $t = 0$, the response of the volume is analogous to a planar semi-infinite body with step-changed surface temperature:

$$\frac{T(r,t) - T_0}{T_w - T_0} = \frac{R}{r} \operatorname{erfc}\left(\frac{r - R}{2\sqrt{\alpha t}}\right) \tag{3.68}$$

For radii such that $R/r \approx 1$, this solution is identical to the planar solution with $x = r - R$.

Cylindrical cases of radial heat flow do not transform neatly into planar cases. However, we may consider the small time behavior of a volume bounded internally by a cylindrical surface of radius R and having an initial temperature of T_0. If the temperature of the surface $r = R$ is stepped to T_w at $t = 0$, the response of the volume for small values of $\alpha t / R^2$ is

$$\frac{T(r,t) - T_0}{T_w - T_0} \approx \left(\frac{R}{r}\right)^{1/2} \operatorname{erfc}\left(\frac{r - R}{2\sqrt{\alpha t}}\right) \tag{3.69}$$

Modeling semi-infinite bodies. Some modeling tactics for semi-infinite bodies include the following:

- Use the semi-infinite solutions directly, when they are applicable.
- Bound complex cases with simpler solutions, for example, by neglecting a thermal resistance or capacitance to get an upper bound on the rate of response or by approximating nonplanar geometries as planar.
- Use an unsteady thermal resistance (see Examples 3.5 and 3.6).
- Time average a periodic unsteady process (see Section 4.5).

Example 3.4 *Initial contact of glass and mold* In Example 3.3, glass at 982°C was brought into contact with a copper mold at 260°C. What is the initial temperature of the interface between them? For how long can this system be considered a semi-infinite body problem?

From Eq. (3.67),

$$\frac{T_i - 982}{260 - 982} = \frac{\sqrt{(k\rho c_p)_{Cu}}}{\sqrt{(k\rho c_p)_{glass}} + \sqrt{(k\rho c_p)_{Cu}}} \tag{3.70a}$$

$$= \frac{\sqrt{(380)(8900)(380)}}{\sqrt{(0.9)(2500)(880)} + \sqrt{(380)(8900)(380)}} \tag{3.70b}$$

$$= 0.96 \tag{3.70c}$$

which gives us $T_i = 287°C$.

The bodies remain semi-infinite until $\delta(t)$ reaches the thickness of one or the other. In this case, copper has a higher thermal diffusivity, and a calculation shows that $\delta_{Cu} = 3.65\sqrt{\alpha_{Cu}t} = t_{Cu}$ after roughly 0.1 s. At that time, $\delta_{glass} \ll t_g$.

Example 3.5 *Simplified solution for convection resistance* The case of a semi-infinite body with convection resistance at the surface (third item in Table 3.4) may be approximated using a simplified model based on thermal resistances. The convection resistance outside the body is just $1/\bar{h}A$; the resistance within the body can be estimated using the value for a semi-infinite body with stepped wall temperature (e.g., as previously estimated in Eq. (3.38) or with the exact result which replaces 3 by π in Eq. (3.38)):

$$R_{cond} = \frac{\sqrt{\pi \alpha t}}{kA} \tag{3.71}$$

The heat flow into the wall, $Q(t)$, can be found for these two resistances in series:

$$Q = \frac{(T_\infty - T_0)}{\frac{1}{\bar{h}A} + \frac{\sqrt{\pi \alpha t}}{kA}} = \frac{\bar{h}A(T_\infty - T_0)}{1 + \frac{\bar{h}\sqrt{\pi \alpha t}}{k}} \tag{3.72a}$$

$$= \frac{Q_0}{1 + \beta\sqrt{\pi}} \tag{3.72b}$$

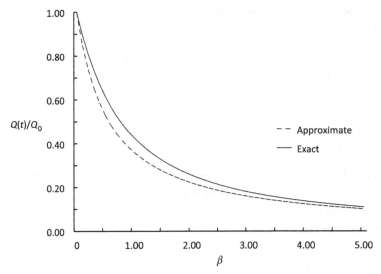

Figure 3.10. Comparison of approximate and exact solutions for a semi-infinite body with a convective boundary condition.

where $Q_0 = \bar{h}A(T_\infty - T_0)$ is the heat flow rate at time 0 and β is defined as before [Eq. (3.65)]. The approximate and exact result are compared in Figure 3.10; the model has a maximum under-predication of about 16% at $\beta \approx 1$.

Unsteady thermal resistances may also be used to derive Eq. (3.67), which gives the interfacial temperature of two semi-infinite bodies that are brought into contact at time zero (see Problem 3.5).

Example 3.6 *Sand Casting* Casting processes involve pouring a molten material into a mold and allowing it to cool until it solidifies. One such process is sand casting, in which molten metals are poured into a cavity in a bed of a sand–clay mixture. The cavity is formed by joining two halves of the sand mold, each of which is created by hydraulically pressing a die into the sand–clay mixture. This process is used, for example, to make cast-iron bathtubs. The molten iron requires about 15 minutes to solidify, after which the sand mold is broken away from the tub.

The primary heat transfer process involved in sand casting is transient conduction of the latent heat of solidification from the metal into the sand. Figure 3.11 shows the situation for a planar casting of surface area A. The molten metal is assumed to be near its melting point temperature, T_{mp}. Some temperature drop may occur through the solidified metal, and additional temperature drop occurs between the surface of the sand mixture, at T_i, and the more distant sand which is at its initial temperature, T_0. Since the sand mixture has a relatively low conductivity, we expect that most of the temperature drop will be in the sand, rather than the solid metal. To estimate the relative magnitudes of these temperature drops, we may consider the thermal circuit shown in Figure 3.12.

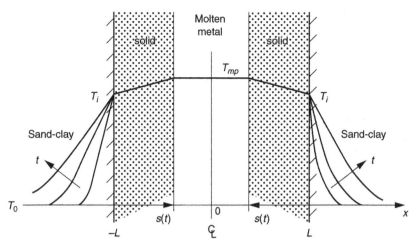

Figure 3.11. Planar sand casting.

This circuit accounts for the conduction resistance of the solidified metal as $(s/k_m A)$, as if steady-state conduction occurred across the metal layer; in other words, it neglects the heat that must be removed when cooling the metal from T_{mp} to T_i, which is acceptable if the sensible heat removed is small relative to the latent heat released during solidification.

The sand mixture is treated differently. Because the sand acts as the sink for latent heat from the metal, its heat capacity must not be neglected. We can approximate the sand as a semi-infinite body for which the surface temperature, T_i, varies little during the solidification process. Then, we can use the heat flux from Table 3.4 to define a time-dependent thermal resistance, as we did in the previous example:

$$R(t) = \frac{(T_i - T_0)}{q_w(t)A} \approx \frac{\sqrt{\pi \alpha_s t}}{k_s A}. \tag{3.73}$$

This time-dependent resistance accounts for both the heat capacity and the conduction resistance of the sand mixture. The numerator in the last term has units of length and is sometimes called a *flux diffusion depth*: $\delta_f = \sqrt{\pi \alpha_s t}$. The flux diffusion depth is about one-half the thermal penetration depth, δ.

Since the heat flow through each resistance in Figure 3.12 is equal,

$$Q = \frac{k_m A(T_{mp} - T_i)}{s(t)} = \frac{k_s A(T_i - T_0)}{\delta_f(t)}. \tag{3.74}$$

Figure 3.12. Thermal circuit for sand casting, using time-dependent thermal resistances.

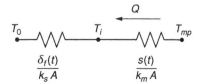

Hence, the relative size of the temperature drop through the metal is

$$\frac{T_{mp} - T_i}{T_i - T_0} = \frac{k_s}{k_m} \frac{s(t)}{\delta_f(t)}. \tag{3.75}$$

The conductivity ratio, k_s/k_m, is likely to be small, since most metals have a conductivity much greater than that of the sand mixture. The size of s/δ_f is less obvious, but we can make an order estimate by using an energy balance: the heat released by solidification from time 0 to t, $\rho_m h_{sf} s(t) A$, must equal that absorbed by the sand, which is approximately $(\rho c_p)_s (T_i - T_0) A \delta/3$ for δ the penetration depth (cf., Eq. 3.32). Thus, with $\delta \approx 2\delta_f$,

$$\frac{\delta_f(t)}{s(t)} = \frac{3\rho_m h_{sf}}{2(\rho c_p)_s (T_i - T_0)} \tag{3.76}$$

$$\geq \frac{3\rho_m h_{sf}}{2(\rho c_p)_s (T_{mp} - T_0)}. \tag{3.77}$$

With Eq. (3.75), we have

$$\frac{T_{mp} - T_i}{T_i - T_0} \leq \frac{k_s}{k_m} \frac{2(\rho c_p)_s (T_{mp} - T_0)}{3\rho_m h_{sf}}. \tag{3.78}$$

Evaluation of this expression for, say, aluminum cast into sand shows the right-hand side to be roughly 6×10^{-4}, verifying that $T_i \approx T_{mp}$ to high accuracy.[7]

At this point, we may simply neglect the presence of the solid metal and directly equate the heat flow into to sand to the rate of heat release during solidification:

$$\frac{k_s A (T_{mp} - T_0)}{\sqrt{\pi \alpha_s t}} = \rho_m h_{sf} A \frac{ds}{dt} \tag{3.79}$$

This differential equation for $s(t)$ may be integrated with $s(0) = 0$ to obtain

$$s(t) = \left(\frac{2k_s}{\rho_m h_{sf}}\right)(T_{mp} - T_0)\sqrt{\frac{t}{\pi \alpha_s}} = \left(\frac{2(\rho c_p)_s (T_{mp} - T_0)}{\rho_m h_{sf}}\right)\sqrt{\frac{\alpha_s t}{\pi}} \tag{3.80}$$

If the half-thickness of the casting is L_c, then the time required to fully freeze the melt, t_f, is given by $s(t_f) = L_c$; substitution of these values into Eq. (3.80) yields

$$t_f = \underbrace{\left[\frac{\rho_m h_{sf}}{2(T_{mp} - T_0)}\right]^2 \frac{\pi}{(k\rho c_p)_s}}_{\equiv\, C,\, \text{Chvorinov's constant}} L_c^2 \tag{3.81}$$

This result is sometimes called *Chvorinov's Rule*. Note that for a given set of materials, $t_f = CL_c^2$, which allows the freezing time to be estimated for geometrically similar molds of a different size. If we define a Fourier number for freezing

---

[7] For aluminum, $T_{mp} = 660°C$, $h_{sf} = 400$ kJ/kg, $k_m \approx 200$ W/m·K, $c_p = 880$ J/kg·K, $\rho_m = 2390$ kg/m³. For a representative foundry sand at these temperatures, $k_s \approx 0.4$ W/m·K, $c_{p,s} \approx 1000$ J/kg·K, $\rho_s \approx 1500$ kg/m³ [3.8].

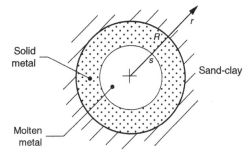

Figure 3.13. Sand casting a spherical object.

as $\mathrm{Fo}_f = \alpha_s t_f / L_c^2$, Chvorinov's Rule may be written as:

$$\frac{\rho_m h_{sf}}{(\rho c_p)_s (T_{mp} - T_0)} = \frac{4}{\sqrt{\pi}} \sqrt{\mathrm{Fo}_f} \tag{3.82}$$

For aluminum cast into foundry sand, this gives $\mathrm{Fo}_f \approx 0.2$.

A few comments should be made. First, if the melt is superheated ($T_{\text{initial}} > T_{mp}$), additional heat must be transferred to the sand before solidification is complete. An estimate of the importance of this additional energy can be made by comparing the latent heat of fusion, h_{sf}, to the sensible heat associated with cooling to the freezing point, $c_{p,m}(T_{\text{initial}} - T_{mp})$. For small superheats, the latent heat will dominate and the melt may be approximated as being at T_{mp} throughout the process. Second, the mold must be sufficiently thick that the semi-infinite body approximation applies up to the time of freezing; this approximation can be checked by comparing $3.65\sqrt{\alpha_s t_f}$ to the mold dimensions. Third, this analysis has neglected any effect of convection in the melt, has assumed that no density change accompanies the phase change, and has assumed that the melt can be treated as a pure substance, with no mushy zone.

Spherical or cylindrical castings: When the object cast is not planar, a greater volume of sand can become involved in the cooling process, reducing the freezing time. For example, suppose that the object cast is nearly spherical, with radius R (Fig. 3.13). Assuming that our other approximations will continue to apply, the wall heat flux can be evaluated from Eq. (3.68):

$$q_w = -k_s \frac{\partial T}{\partial r}\bigg|_{r=R} = k_s(T_{mp} - T_0)\left(\frac{1}{R} + \frac{1}{\sqrt{\pi \alpha_s t}}\right) \tag{3.83}$$

where k_s and α_s are the conductivity and diffusivity of the sand, as before. A heat balance at the surface of the liquid portion of the sphere, neglecting the thermal resistance of the solid part, gives

$$4\pi R^2 \left[k_s(T_{mp} - T_0)\left(\frac{1}{R} + \frac{1}{\sqrt{\pi \alpha_s t}}\right)\right] = \rho_m h_{sf} \frac{4\pi}{3}\frac{d}{dt}(a^3 - s^3) \tag{3.84}$$

for $s(t)$ the radius of the liquid core. Integration from $s(0) = R$ to $s(t_f) = 0$ gives an implicit relationship for t_f:

$$R = \frac{3(T_{mp} - T_0)}{\rho_m h_{sf}} \left(\frac{2}{\sqrt{\pi}} \sqrt{(k\rho c_p)_s t_f} + \frac{k_s t_f}{R} \right) \tag{3.85}$$

If we define a Fourier number for freezing as $\text{Fo}_f = \alpha_s t_f / R^2$, this result may be rewritten as

$$\frac{\rho_m h_{sf}}{(\rho c_p)_s (T_{mp} - T_0)} = \frac{6}{\sqrt{\pi}} \sqrt{\text{Fo}_f} + 3\text{Fo}_f \tag{3.86}$$

Using the properties of aluminum and sand given previously, we find $\text{Fo}_f \approx 0.06$.

An analogous relationship for cylindrical castings may be derived from Eq. (3.69), but it will be restricted to small values of $\alpha_s t_f / R^2$ ($\lesssim 0.3$ or so).

Example 3.7 *Vapor bubble growth* When vapor bubbles form within a superheated liquid, heat is conducted from the hot liquid at T_{sup} to the liquid–vapor interface at T_{sat} providing energy for vaporization of the liquid. Thus, these bubbles grow rapidly once they are large enough to overcome the restraining effects of surface tension and liquid inertia.[8] Time scales for this process are typically measured in milliseconds.

The rate of growth of such bubbles is determined by the rate of heat transfer from the bulk liquid to the surface of the bubble; and, because the bubbles are typically small with respect the dimensions of the liquid pool (on the order of a millimeter in diameter, say), we may regard the process as heat transfer in a semi-infinite volume bounded internally by the spherical surface of the bubble, which has radius $a(t)$. A first cut at this problem might use the derivative of Eq. (3.68), for transient conduction outside a sphere, to calculate the heat flux to the liquid–vapor interface, which should equal the rate at which latent heat is taken up by vapor formation. The energy balance is

$$4\pi a^2 k_l \left. \frac{\partial T}{\partial r} \right|_{r=a} = \rho_v h_{fg} \frac{d}{dt} \left(\frac{4\pi a^3}{3} \right) \tag{3.87}$$

Differentiating Eq. (3.68) gives

$$\left. \frac{\partial T}{\partial r} \right|_{r=a} = -(T_{sat} - T_{sup}) \left(\frac{1}{a} + \frac{1}{\sqrt{\pi \alpha_l t}} \right) \tag{3.88}$$

The term $1/a$ on the left-hand side represents the effect of curvature; had we used the planar semi-infinite conduction solution, that term would be absent from this equation. Upon substituting Eq. (3.88) into Eq. (3.87), we have

$$k_l (T_{sup} - T_{sat}) \left(\frac{1}{a} + \frac{1}{\sqrt{\pi \alpha_l t}} \right) = \rho_v h_{fg} \frac{da}{dt} \tag{3.89}$$

---

[8] For details, see L.E. Scriven [3.9] or L.S. Tong [3.10, Chapter 2].

As we saw for sand casting in Example 3.6, interfaces in semi-infinite problems tend to grow as $\sqrt{\alpha t}$, so we might try assuming that

$$a(t) = 2C\sqrt{\alpha_l t} \tag{3.90}$$

for an as yet unknown C. Substitution of this equation into Eq. (3.89) and some rearrangement gives us an implicit expression for C

$$C^2 = \left(\frac{\mathrm{Ja}\,\rho_l}{\rho_v}\right)\left[\frac{1}{2} + \frac{C}{\sqrt{\pi}}\right] \tag{3.91}$$

where we define the *Jakob number*, Ja, as

$$\mathrm{Ja} \equiv \frac{c_p(T_{\mathrm{sup}} - T_{\mathrm{sat}})}{h_{\mathrm{fg}}} \tag{3.92}$$

The Jakob number compares the sensible heat of the liquid for the given super-heat to the latent heat of vaporization. At pressures low compared to the critical pressure, $\rho_l/\rho_v \gg 1$; and for typical values of Ja, the group $(\mathrm{Ja}\,\rho_l/\rho_v)$ is large compared to one, so that C is a large number. In this case, the term $1/2$ on the right-hand side of Eq. (3.91), which results from the curvature of the interface, is negligible; and C is given approximately by

$$C \approx \left(\frac{\mathrm{Ja}\,\rho_l}{\rho_v\sqrt{\pi}}\right) \tag{3.93}$$

This result was first derived by Max Jakob in the 1930s using the *planar* solution.

Unfortunately, the rate of bubble growth predicted by the preceding analysis is low by almost a factor of two. Our first suspicion might be that the neglected curvature effect is the problem; however, this is not the case. The diffusion layer around the bubble is small relative to the bubble radius. For example, depending on the superheats involved and the nucleation process considered, a steam bubble might have a radius of 0.7 mm after growing for 10 ms. The diffusion length, which is the penetration of the temperature reduction into the liquid around the bubble, would be roughly $3.65\sqrt{\alpha_l t} = 0.15$ mm, too small for curvature cause such a large error.

In fact, the problem is that the expansion of the bubble creates a radial velocity gradient in the liquid near the bubble surface. As a result of the density difference between the liquid and the vapor, the growth of the bubble displaces the liquid radially. Mass conservation applied to a control volume of radius $r > a$ allows us to determine the liquid's radial velocity $u(r, t)$:

$$\frac{d}{dt}\left(\rho_v\frac{4\pi a^3}{3} + \rho_l\frac{4\pi\left(r^3 - a^3\right)}{3}\right) = -4\pi r^2\rho_l\,u(r, t) \tag{3.94}$$

From this, we find the liquid velocity in terms of the interface speed, $\dot{a} = da/dt$:

$$u(r, t) = \left(\frac{\rho_l - \rho_v}{\rho_l}\right)\dot{a}\frac{a^2}{r^2} \approx \dot{a}\frac{a^2}{r^2} \tag{3.95}$$

where the final approximation is valid at pressures low compared to the critical pressure. At the bubble surface, the liquid speed is equal to the interface speed, so that the relative velocity between the liquid and the interface is 0. Farther from the interface, however, the liquid moves more slowly, and the net effect is to bring hotter liquid toward the bubble, steepening the radial temperature gradient and increasing the vaporization rate.

As we have already seen, curvature is not important to the temperature profile, so we can write the energy equation in a planar form, but with the addition of a convective term to address the liquid motion. It is convenient to use a reference frame that moves with the surface of the bubble and a coordinate $x = r - a(t)$. The energy equation including the liquid velocity in the radial direction is

$$\frac{\partial T}{\partial t} + u_x \frac{\partial T}{\partial x} = \alpha_l \frac{\partial^2 T}{\partial x^2}. \tag{3.96}$$

In the vicinity of the interface,

$$\frac{a^2}{r^2} = \frac{a^2}{(a+x)^2} \approx 1 - \frac{2x}{a} + \cdots$$

so that the velocity of the liquid with respect to the interface is

$$u_x(x,t) = u(r,t) - \dot{a}(t) \approx -\dot{a}(t) \frac{2x}{a}. \tag{3.97}$$

The energy equation becomes

$$\frac{\partial T}{\partial t} - \dot{a} \frac{2x}{a} \frac{\partial T}{\partial x} = \alpha_l \frac{\partial^2 T}{\partial x^2}. \tag{3.98}$$

The only length scale in this problem is the bubble radius, which we shall continue to assume is of the form $a(t) = 2C\sqrt{\alpha_l t}$. Further, because the length scale is a function of time, the situation suggests a similarity solution. Therefore, we apply the usual similarity coordinate

$$s = \frac{x}{2\sqrt{\alpha_l t}} \tag{3.99}$$

to the energy equation (3.98) and, after a bit of algebra, obtain

$$\frac{d^2 T}{ds^2} + \sqrt{3}s \frac{dT}{ds} = 0 \tag{3.100}$$

This equation is easily integrated by reduction of order, and with the two boundary conditions, $T(x = 0) = T_{\text{sat}}$ and $T(x \to \infty) = T_{\text{sup}}$, becomes

$$T(x,t) = T_{\text{sup}} - (T_{\text{sup}} - T_{\text{sat}}) \, \text{erfc}\left(\frac{\sqrt{3}x}{2\sqrt{\alpha_l t}}\right). \tag{3.101}$$

Substitution of this temperature profile and the expression for $a(t)$ into the interfacial energy balance (Eq. 3.87) leads to the value of C:

$$\rho_v h_{fg} \frac{da}{dt} = k_l \left. \frac{\partial T}{\partial x} \right|_{x=0} \tag{3.102}$$

$$\rho_v h_{fg} C \sqrt{\frac{\alpha_l}{t}} = k_l (T_{sup} - T_{sat}) \left(\frac{2e^{-0^2}}{\sqrt{\pi}} \frac{\sqrt{3}}{2\sqrt{\alpha_l t}} \right)$$

$$C = \sqrt{\frac{3}{\pi} \frac{Ja\,\rho_l}{\rho_v}}. \tag{3.103}$$

Thus, the bubble radius may be computed as

$$a(t) = 2\sqrt{\frac{3}{\pi} \left(\frac{Ja\,\rho_l}{\rho_v} \right)} \sqrt{\alpha_l t}. \tag{3.104}$$

This result is in good agreement with measurements [3.11].

3.5 Nondimensionalization and Scaling in Transient Conduction

We considered nondimensionalization of steady conduction problems in Section 2.6. We now wish to extend these ideas to unsteady conduction problems. Key differences include the appearance of characteristic time scales, the possibility that characteristic temperatures vary in time, and the role of propagation of temperature changes through the system.

Typically, the time scale is found by balancing the effects of heat capacitance and thermal resistance in the system at hand. For distributed heat conduction problems, the storage and conduction terms in the heat equation, Eq. (3.42), may be compared to find the time scale. In the case of systems that are tied to thermal resistances external to the body through which heat flows (a hot metal billet cooling in air, for instance), the time scale is found by balancing the internal heat storage against the internal and external thermal resistances. Occasionally, a time-dependent heat source will define the time scale, as for example when daily temperature variation leads to periodic heating and cooling of a system.

3.5.1 Characteristic Temperatures

Temperature changes generally scale with the maximum temperature difference in the problem,

$$\Delta T = T_{max} - T_{min} \tag{3.105}$$

In unsteady cooling problems, the maximum temperature will generally be the initial temperature and the minimum temperature will be the temperature of the final steady state, such as the temperature of the medium that cools the object. The final steady temperature of a system can usually be determined either by inspection, if the system runs down to thermal equilibrium with its environment, or by separately solving the steady-state conduction problem that characterizes the system after the transient temperature change has passed.

Once ΔT is found, a characteristic dimensionless temperature is formed as usual by dividing the difference between local temperature and a minimum (or maximum) temperature by ΔT

$$\Theta = \frac{T(x, y, z, t) - T_{min}}{T_{max} - T_{min}} \tag{3.106}$$

Under this scaling, Θ will vary between 0 and 1, which simply tells you that during the cooling process the temperature is always going to vary between its minimum and maximum values.

If the maximum and minimum temperatures are independent of position and time (and they will be if they are chosen properly), the heat equation can be rewritten in terms of Θ

$$\rho c_p \frac{\partial \Theta}{\partial t} = k \nabla^2 \Theta + \frac{\dot{q}_v}{\Delta T} \tag{3.107}$$

In cases where volumetric heating is present – for example, the initial warm-up of a nuclear fuel rod that is convectively cooled by pressurized water – the maximum temperature of the problem must normally be found by solving for the final steady-state maximum temperature. In such cases, ΔT will normally be linearly proportional to $\dot{q}_v$.

3.5.2 Characteristic Lengths

Any finite-sized body has at least one characteristic length scale. Depending on whether the thermal resistance is mainly external (low Bi) or also internal, we may approach the length scale differently. In the low Biot number case, when a lumped capacity model applies, the length scale will appear as a volume-to-surface area ratio in the lumped time constant [Eq. (3.12)], where the volume relates to the heat capacitance of the object and the surface area rates to the external thermal resistance.

When internal conduction resistance is important, the direction of heat flow in the body is also significant; and, as we saw in Section 2.6, the characteristic length scale may be different for heat flow in different directions. For example, when a thin slab is cooled through its surface, the characteristic dimension for heat flow through the thickness of the slab is much less than that for heat flow in the plane of the slab. If the plate is being cooled through the plane surfaces, the resistance through the thickness direction will govern the rate of cooling. For example, if a thin plate of thickness d and width and length L is cooled through its surface, the characteristic length across which heat is conducted in the thickness direction (x) is $d/2$, while that in the length/width directions (y, z) is $L/2$. Since $L \gg d$, this means that the temperature gradients will be greater in the thickness direction, or that conduction resistance will be lower in the thickness direction, than in the other directions. We can show this by scaling the heat equation. We introduce scaled coordinates

$$\hat{x} = x/(d/2) \tag{3.108a}$$

$$\hat{y} = y/(L/2) \tag{3.108b}$$

$$\hat{z} = z/(L/2) \tag{3.108c}$$

so that the Laplacian becomes

$$\nabla^2 T = \Delta T \left(\frac{\partial^2 \Theta}{\partial x^2} + \frac{\partial^2 \Theta}{\partial y^2} + \frac{\partial^2 \Theta}{\partial z^2} \right)$$

$$= \Delta T \left(\frac{4}{d^2} \frac{\partial^2 \Theta}{\partial \hat{x}^2} + \frac{4}{L^2} \frac{\partial^2 \Theta}{\partial \hat{y}^2} + \frac{4}{L^2} \frac{\partial^2 \Theta}{\partial \hat{z}^2} \right) \tag{3.109}$$

$$= \left(\frac{4\Delta T}{d^2} \right) \left[\frac{\partial^2 \Theta}{\partial \hat{x}^2} + \left(\frac{d}{L} \right)^2 \frac{\partial^2 \Theta}{\partial \hat{y}^2} + \left(\frac{d}{L} \right)^2 \frac{\partial^2 \Theta}{\partial \hat{z}^2} \right]$$

Now, if the length and temperature scales have been chosen correctly all of the nondimensional derivatives are of order 1 (orders of magnitude were introduced in Section 2.6):

$$\frac{\partial^2 \Theta}{\partial \hat{x}^2} = \mathcal{O}(1) \tag{3.110a}$$

$$\frac{\partial^2 \Theta}{\partial \hat{y}^2} = \mathcal{O}(1) \tag{3.110b}$$

$$\frac{\partial^2 \Theta}{\partial \hat{z}^2} = \mathcal{O}(1) \tag{3.110c}$$

Since the plate is thin, $d \ll L$. Thus,

$$\nabla^2 T = \left(\frac{4\Delta T}{d^2} \right) \Bigg[\underbrace{\frac{\partial^2 \Theta}{\partial \hat{x}^2}}_{\mathcal{O}(1)} + \underbrace{\left(\frac{d}{L} \right)^2}_{\ll 1} \underbrace{\frac{\partial^2 \Theta}{\partial \hat{y}^2}}_{\mathcal{O}(1)} + \underbrace{\left(\frac{d}{L} \right)^2}_{\ll 1} \underbrace{\frac{\partial^2 \Theta}{\partial \hat{z}^2}}_{\mathcal{O}(1)} \Bigg] \tag{3.111}$$

$$\approx \left(\frac{4\Delta T}{d^2} \right) \left(\frac{\partial^2 \Theta}{\partial \hat{x}^2} \right)$$

The Laplacian is thus on the order of $4\Delta T/d^2$. The y and z components of the Laplacian are negligible because the conduction heat flux in the y and z directions is negligible in comparison to the heat flux in the x direction. (Near the corners of the plate, the length scale for conduction in all three directions will be of the same order; and a different scaling should be applied if a precise analysis of the corners is desired.)

It is sometimes useful to seek a single length scale for an irregular object, when only a crude estimate is desired, and one approach is dividing an object's the volume, $\mathcal{V}$, by the surface area through which heat flows, $\mathcal{A}$:

$$\mathcal{L} = \frac{\mathcal{V}}{\mathcal{A}} \tag{3.112}$$

For example, for a cube having sides of length L cooled through all six surfaces,

$$\mathcal{L} = \frac{L^3}{6L^2} = \frac{L}{6} \tag{3.113}$$

meaning that the heat generally flows only a distance of $L/6$ to reach the surface. In this three-dimensional situation, most of the heat has only a short distance to travel when leaving the cube, especially in the regions near the corners. Heat leaving the center of the cube will travel a much greater distance — between $L/2$ and $\sqrt{3}\,L/2$. Thus, the volume-to-area length scale does not differentiate between the cooling of the surface of a body and the cooling of the interior. Estimates of the cooling time based on such a length scale will consequently be averages for the process.

3.5.3 Time Scales for Situations Having a Length Scale

When the problem has a length scale, we can estimate characteristic response times by scaling the heat equation. When no set length scale is present, as for a semi-infinite body, the time scale may be that required for a temperature change to travel some particular distance into an object; and, as discussed in Section 3.2.3, the length and time scales are not independent.

Suppose that we wish to find the time scale over which the thin plate of thickness d described by Eq. (3.111) cools. Let us introduce an unknown conduction time scale, t_d, and scale time as

$$\hat{t} = \frac{t}{t_d} \tag{3.114}$$

If t_d represents the characteristic time for cooling by conduction, then $\hat{t}$ is of order one when the plate has cooled. By comparison to Eq. (3.41), we see that $\hat{t}$ is a Fourier number. Practically speaking, the value of $\hat{t}$ may be greater or less than one when we say that the plate is "cold" – as we saw in Section 3.2, the cooling process has a long exponential tail which must be arbitrarily truncated.

Introducing the time scale t_d into the partially nondimensionalized heat equation, Eq. (3.107), together with our scaled Laplacian, Eq. (3.111), we have (assuming that $\dot{q}_v = 0$)

$$\underbrace{\frac{\rho c_p \Delta T}{t_d} \frac{\partial \Theta}{\partial \hat{t}}}_{\mathcal{O}(1)} = k\Delta T \, \nabla^2 \Theta$$

$$\approx k \left(\frac{4\Delta T}{d^2} \right) \underbrace{\left(\frac{\partial^2 \Theta}{\partial \hat{x}^2} \right)}_{\mathcal{O}(1)} \tag{3.115}$$

Since the two sides of the equation must be of the same order of magnitude, it follows that

$$\frac{\rho c_p}{t_d} = \mathcal{O}\left(\frac{4k}{d^2} \right) \tag{3.116}$$

or

$$t_d = \mathcal{O}\left(\frac{\rho c_p d^2}{4k}\right) = \mathcal{O}\left(\frac{d^2}{4\alpha}\right) \tag{3.117}$$

This is identical to the diffusional time scale we discussed in Section 3.2.4, for a plate of thickness $d = 2L$.

The scaling analysis we have just completed shows that the time for a plate to cool is on the order of t_d. In Section 3.2, we analyzed the rate process more carefully using thermal resistances and capacitances to calculate the temperature response, finding that the Fourier number at which the cooling process is 95% complete is $t/t_d = 3/2$ when $\mathrm{Bi} \gg 1$. Thus, the present estimate of the "cooling time" is lower by $3/2$. In addition, if we were to deal more carefully with the boundary conditions in this scaling estimate, we would expect the Biot number to appear in the cooling time, at least when the Biot number is small.

It should be mentioned that if the conduction length scale were chosen differently (e.g., as $d/4$ for a plate of thickness d) that t_d would be different (e.g., ¼ the value just found). This ambiguity occurs because our estimate is only an order of magnitude calculation.

Example 3.8 *Scaling of an unsteady fin* Suppose that a long thermocouple probe extends from a wall into an airstream. Initially, the system is cold and isothermal at T_0. The temperature of the airstream rises abruptly to T_∞ at time 0, causing both the probe and the wall to begin to warm up. The wall is massive, and its temperature increases relatively slowly. The probe, on the other hand, is small and should nearly track the gas temperature, except for heat conduction along its axis into the wall. When steady state is reached, the probe will behave like a fin, with its base at the wall temperature and its tip nearer to the gas temperature. Before that, the probe behaves as an *unsteady* fin. Let us determine appropriate simplified equations for the temperature response of the probe tip, assuming that conditions are as described in Problem 3.6 and ignoring thermal radiation.

The steady fin equation was developed in Section 2.5. Here we use the unsteady fin equation, for temperature $T(x, t)$, which includes a term representing the rate at which energy is stored by the thermal capacity of the fin at each position:

$$\underbrace{\rho c_p A_c \frac{\partial T}{\partial t}}_{\text{storage}} - \underbrace{kA_c \frac{\partial^2 T}{\partial x^2}}_{\text{conduction}} + \underbrace{\bar{h}\mathcal{P}(T - T_\infty)}_{\text{convection}} = 0 \tag{3.118}$$

To scale this equation, we first nondimensionalize temperature using the initial temperature of the system, T_0, and the new temperature of the gas, T_∞:

$$\Theta \equiv \frac{T(x, t) - T_\infty}{T_0 - T_\infty} \tag{3.119}$$

The scaled temperature, Θ, clearly varies between 0 and 1.

We scale x and t with characteristic dimensions, x_c and t_c, without specifying yet what these dimensions are

$$\hat{x} \equiv \frac{x}{x_c} \tag{3.120}$$

$$\hat{t} \equiv \frac{t}{t_c} \tag{3.121}$$

With these scalings, the unsteady fin equation becomes

$$\left(\frac{\rho c_p A_c}{\bar{h}P}\right) \frac{1}{t_c} \frac{\partial \Theta}{\partial \hat{t}} - \left(\frac{kA_c}{\bar{h}P}\right) \frac{1}{x_c^2} \frac{\partial^2 \Theta}{\partial \hat{x}^2} + \Theta = 0 \tag{3.122}$$

We may identify the groups in parentheses as the lumped capacity time constant, τ, and the fin parameter, m^2:

$$\tau \equiv \left(\frac{\rho c_p A_c}{\bar{h}P}\right) \tag{3.123}$$

$$m^2 \equiv \left(\frac{\bar{h}P}{kA_c}\right) \tag{3.124}$$

Do these two parameters give the relevant characteristic time and length for the problem? Note the probe length, L, is not present at this point! It appears that the length scale of the fin parameter, $1/m$, is the conduction length scale in this case. In fact, if the numbers from Problem 3.6 are used, $1/m = 1.3$ cm while $L = 30$ cm. This suggests that heat conduction will be limited to a short length of the probe near the wall, a distance that our knowledge of steady fins suggests should be on the order of $3/m$. The remainder of the probe will be unaffected by conduction to the wall, and so should have a relatively uniform temperature. The probe's time response may be different in each of these regions. The time constant τ, on the other hand, gives the scale of the time interval during which temperature changes at all locations in the probe.

The scaled equation is

$$\left(\frac{\tau}{t_c}\right) \frac{\partial \Theta}{\partial \hat{t}} - \left(\frac{1}{m^2 x_c^2}\right) \frac{\partial^2 \Theta}{\partial \hat{x}^2} + \Theta = 0 \tag{3.125}$$

The selection of x_c and t_c is a matter of deciding for what intervals of time or position we wish to obtain an solution. For example, if we consider times that are large compared to the time constant ($t_c \gg \tau$), the time derivative will be negligible and we are left with the steady-state solution. Considering $x_c = L \gg 1/m$ will eliminate the conduction term: this implies that a substantial portion of the probe is not affected by conduction to the wall (Fig. 3.14). From such considerations, we can identify three relevant cases.

Case 1.

$$\left(\frac{\tau}{t_c}\right) = \mathcal{O}(1) \quad \text{with} \quad \left(\frac{1}{m^2 x_c^2}\right) \ll 1$$

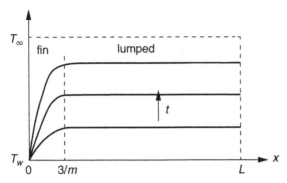

Figure 3.14. Temperature distribution in an unsteady fin.

This situation would correspond to selecting the characteristic time to be $t_c = \tau$ and considering characteristic lengths $x_c \gg 1/m$ (e.g., $x_c = L$). The differential equation simplifies to

$$\frac{d\Theta}{d\hat{t}} + \Theta = 0 \tag{3.126}$$

which is simply lumped capacity response. This solution applies to most of the probe, away from the end that contacts the wall.

Case 2.

$$\left(\frac{\tau}{t_c}\right) \ll 1 \quad \text{with} \quad \left(\frac{1}{m^2 x_c^2}\right) = \mathcal{O}(1)$$

This situation would correspond to selecting the characteristic time to be $t_c \gg \tau$, much longer than the transient response time, and considering characteristic lengths $x_c = 1/m$. The differential equation simplifies to

$$-\frac{d^2\Theta}{d\hat{x}^2} + \Theta = 0 \tag{3.127}$$

which is the nondimensional steady fin equation. This equation applies for times large compared to the time constant. Heat is conducted from the tip of the probe at T_∞ (in steady state) to the base of the probe at T_w, but most of the temperature variation will be in the region $x < 3/m$.

Case 3.

$$\left(\frac{\tau}{t_c}\right) = \mathcal{O}(1) \quad \text{with} \quad \left(\frac{1}{m^2 x_c^2}\right) = \mathcal{O}(1)$$

This situation would correspond to selecting the characteristic time to be $t_c = \tau$ and considering characteristic lengths $x_c = 1/m$. The differential equation to be solved is the unsteady fin equation,

$$\frac{\partial\Theta}{\partial\hat{t}} - \frac{\partial^2\Theta}{\partial\hat{x}^2} + \Theta = 0. \tag{3.128}$$

This equation describes the transient response of the probe in the region near the wall. In this region, heat is conducted from the tip of the probe, which has a rising temperature, to the base of the probe at T_w.[9]

REFERENCES

[3.1] M. F. Ashby. *Materials Selection in Mechanical Design*. Oxford: Butterworth-Heinemann, 1992.

[3.2] A. G. Ostrogorsky. Simple explicit equations for transient heat conduction in finite solids. *Journal of Heat Transfer*, **131**, 011303, 2009.

[3.3] J. H. Lienhard IV, and J. H. Lienhard V. *A Heat Transfer Textbook*, 4th ed. Mineola, NY: Dover Publications, 2011. http://ahtt.mit.edu.

[3.4] H. S. Carslaw, and J. C. Jaeger. *Conduction of Heat in Solids*, 2nd ed. New York: Oxford University Press, 1959.

[3.5] P. J. Schneider. *Temperature Response Charts*. New York: John Wiley & Sons, 1963.

[3.6] H. D. Baehr, and K. Stephan. *Heat and Mass Transfer*. Berlin: Springer-Verlag, 1998.

[3.7] M. Hambrey, and J. Alean. *Glaciers*. New York: Cambridge University Press, 1992.

[3.8] G. Malherbe, J.-F. Henry, C. Bissieux, A. El Bakali, and S. Fohanno. Measurement of thermal conductivity of granular materials over a wide range of temperatures. comparison with theoretical models. *Journal of Physics: Conference Series*, **395**, 012081, 2012.

[3.9] L. E. Scriven. On the dynamics of phase growth. *Chemical Engineering Science*, **10**, 1–13, 1959.

[3.10] L. S. Tong. *Boiling Heat Transfer and Two-phase Flow*. New York: John Wiley & Sons, 1965.

[3.11] P. Dergarabedian. The rate of growth of vapor bubbles in superheated water. *Journal of Applied Mechanics, Transactions of ASME*, **75**, 537–545, 1953.

[3.12] V. S. Arpaci. *Conduction Heat Transfer*. Needham Heights, MA: Ginn Press/Pearson Custom Publishing, 1991.

[3.13] F. Kreith. *Principles of Heat Transfer*, 2nd ed. Scranton, PA: International Textbook Company, 1965.

[3.14] P. B. Whalley. *Boiling, Condensation, and Gas-Liquid Flow*. Oxford: Oxford University Press, 1987.

[3.15] A. Yamanouchi. Effect of core spray cooling in transient state after loss of coolant accident. *Journal of Nuclear Science and Technology*, **5**(11), 547–558, 1968.

[3.16] K. H. Sun, G. E. Dix, and C. L. Tien. Cooling of a very hot vertical surface by a falling liquid film. *Journal of Heat Transfer*, **96**(2), 126–131, 1974.

---

[9] Comparing Eqs. (3.128) and (3.126), we see that when the parameter $1/m^2 x_c^2$ is small (for $x_c = L$), the highest derivative in the equation is lost. The solution of the equation without this derivative (the lumped solution) cannot satisfy the boundary condition at the wall, $T(x = 0) = T_w$. Mathematically, this is a *singular perturbation problem*, in which the small size of a parameter leads to the disappearance of a higher order derivative in the differential equation. Consequently, the derivative cannot be neglected in a thin region near the wall, for $x_c = \mathcal{O}(1/m)$, and a different "inner" solution, Eq. (3.128), is required there. For the purpose of our original problem, it is sufficient to use only the "outer" solution, Eq. (3.126), which describes the temperature response of the probe tip.

PROBLEMS

3.1 To monitor the temperature of a hot gas flowing through a stainless steel pipe, an RTD sensor is epoxied to the outside surface of the pipe (Fig. 3.15). The pipe and sensor sit in still air at $T_\infty = 25°C$. Initially, the gas flowing through the pipe is also at 25°C, but at some time its temperature abruptly rises to 100°C. In this problem, we would like you to estimate how quickly the RTD sensor responds to the gas temperature change.

You may treat the sensor as a homogeneous cylinder having diameter $D_s = 6$ mm and length $L_s = 50$ mm with $k_s = 10$ W/m·K and $(\rho c_p)_s = 3.5 \times 10^6$ J/m³K. The pipe has an outside diameter of $D_p = 11.4$ cm and a wall thickness of $t_p = 6$ mm; for the pipe, $k_p = 14$ W/m·K and $(\rho c_p)_p = 3.2 \times 10^6$ J/m³K. The epoxy has a mean thickness of $t_e = 2$ mm with $k_e = 0.3$ W/m·K and $(\rho c_p)_e = 2.0 \times 10^6$ J/m³K.

The heat transfer coefficient to the gas inside the tube is $\bar{h}_i = 150$ W/m²·K. The effective heat transfer coefficient for convection and radiation outside the tube is $\bar{h}_o = 20$ W/m²·K.

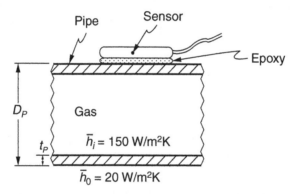

Figure 3.15. Pipe wall with RTD sensor epoxied to it.

a. Estimate the time for the pipe wall temperature, T_w, to respond to the change in gas temperature, T_g, if there is no sensor attached to the pipe.

b. Represent the pipe and sensor using a thermal circuit. Neglect heat conduction in the axial (z) and azimuthal (θ) directions in the pipe wall. Calculate the values of each resistance and capacitance in your circuit. Do the values suggest any possible simplification of the circuit?

c. Your simplified circuit can be represented with coupled first-order equations for the temperatures. Write these equations. If you are comfortable with simulation software, you may solve the equations numerically. Plot your results.

d. Discuss qualitatively the effect on time response of axial and azimuthal conduction in the pipe, if they are not negligible.

3.2 In Eq. (3.31), a parabolic model was used for the unsteady temperature profile. In this problem, we consider two alternative models for the profile.

a. Repeat the calculations of U, R_{cond}, δ, and q_w using a linear profile, $f(x) = 1 - x/\delta$.

b. With a cubic temperature profile, we can satisfy one more boundary condition. Use the heat equation to show that such a boundary condition is $f''(0) = 0$. Then construct an appropriate cubic profile and repeat the calculations of U, R_{cond}, δ, and q_w.

c. Compare and discuss the results for q_w and δ given by these profiles.

3.3 In this problem, we would like you to evaluate the capabilities of the plastic melting system illustrated in Figure 3.16. A thermosetting plastic powder is forced through a heater which consists of a 30 mm OD block of stainless steel with seven parallel holes (6 mm ID) drilled through it. The block of metal is heated to 150°C (by an electric heater wrapped over its surface), and the heat is transferred to the plastic, which enters at 20°C. The properties of the plastic are given below and in Figure 3.17.

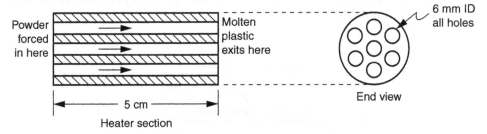

Figure 3.16. Configuration of a plastic melter.

a. How does the temperature distribution in the plastic compare to that in the steel? What can we conclude about the temperature along the walls of the holes? Is axial or radial heat conduction more significant in the plastic?

b. What can we conclude about the velocity distribution in the plastic as it moves through the passages?

c. Make an upper bound estimate of the rate (in m/hr) at which the plastic can be pushed through the heater if all of it is to be melted.

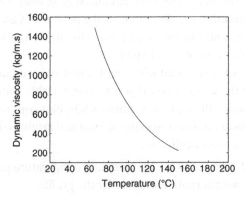

Figure 3.17. Viscosity of molten plastic.

Properties of Plastic (average from 20 to 150°C)

melting point	$T_{mp} = 66°C$
thermal conductivity	$k = 0.23$ W/m·K
specific heat capacity	$c_p = 2300$ J/kg·K
density	$\rho = 1280$ kg/m³
thermal diffusivity	$\alpha = 7.8 \times 10^{-8}$ m²/s

3.4 The following item ran in a New England newspaper several years ago. Explain what is responsible for the phenomenon described.

Temperatures Rise; So Do Dangers
Plumbers say Warmth Can Burst Pipes

Temperatures may rise in the next couple of days, but a double-digit reading doesn't mean the danger of frozen pipes has passed.

Among plumbers, it's common knowledge that during warm weather that follows a cold snap, water lines and indoor pipes freeze and break.

When outdoor temperatures rise after a period of freezing weather, the cold in the ground or an exterior wall of a building is pushed inward and can freeze pipes that are already borderline, said a local plumbing and heating contractor.

"It's sort of a 'straw that breaks the camel's back' situation," he said. "The first 24 to 48 hours, warmer weather just drives the cold right in.

"We get a lot of calls right at the beginning of a snap and then again right afterwards," he added.

3.5 Use an unsteady thermal resistance model to derive an expression for the temperature at the interface of two semi-infinite bodies at initial temperatures T_1 and T_2 which are brought into contact at time 0. Also derive an expression for the heat flux between them.

3.6 A power plant chimney is constructed of firebrick and has an internal diameter of 1.0 m and a wall thickness of 0.2 m. The chimney is isothermal at 20°C after a shutdown. When the power plant restarts, exhaust (mainly air) enters the bottom of the chimney at a speed of 20 m/s and a temperature of 700°C. The pressure in the chimney is approximately 1 atm.

At a location 15 m above the chimney's inlet, two thermocouples have been placed. One is mounted flush against the chimney surface. The second lies at the tip of 0.3 m long stainless steel thermocouple well (6 mm OD) that reaches into the exhaust stream. Dimensions are shown in Figure 3.18 and thermophysical properties are listed below.

One hour after the power plant restarts, obtain upper and lower bounds on the following temperatures:

- The bulk temperature of the air passing the thermocouples
- The temperature of the chimney wall at this location
- The temperature of the thermocouple in the thermowell

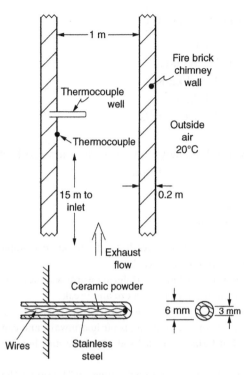

Figure 3.18. Chimney and thermocouple well.

Properties of Stainless Steel

thermal conductivity	$k = 25$ W/m·K
specific heat capacity	$c_p = 400$ J/kg·K
density	$\rho = 8000$ kg/m^3
total emissivity	$\varepsilon \approx 0.5$

Properties of Firebrick

thermal conductivity[10]	$k = 0.1$ W/m·K
specific heat capacity	$c_p = 960$ J/kg·K
density	$\rho = 2000$ kg/m^3
total emissivity	$\varepsilon \approx 0.59$

3.7 A thin metal plate is cooled by laying it on an insulating surface and blowing air across the top of it (Fig. 3.19). The plate initially has a uniform temperature, $T_0 = 1000°$C, and the air flow has a freestream temperature $T_\infty = 27°$C and speed $u_\infty = 10$ m/s. The plate is $L = 60$ cm long and $H = 8$ mm thick. The plate is left to cool until no part of it is above $200°$C.

Owing to the axial growth of the boundary layer, the forward part of the plate may cool faster than the rear part. The plate can be considered to be made from any of a variety of common metals.

---

[10] Refractory bricks are available in many grades. The conductivity listed here is for an insulating brick. Common fireclay bricks have a thermal conductivity in the range of 0.7 to 1.6 W/m·K.

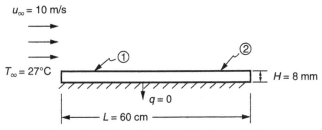

Figure 3.19. Metal plate cooling in an air flow.

a. For a given air speed and plate length, what process is mainly responsible for determining whether there is a temperature gradient along the axis of the plate?

b. What conclusions can you draw about temperature gradients through the thickness of a metal plate having these dimensions and cooled at this airspeed?

c. Consider a location 1 that is 12.5 cm from the leading edge of the plate and a location 2 that is 50.0 cm from the leading edge. Obtain an upper bound on the temperature difference between those two points as a function of time t.

d. Determine a dimensionless group whose size will indicate whether axial temperature gradients are significant.

3.8 A fine tungsten wire is to be used as a filament in an incandescent light bulb. An AC electrical current is passed through the wire, heating it to a high temperature. The filament then radiates heat (some of it in visible wavelengths) from its surface. The filament sits inside a large glass bulb that is filled with inert gas at a temperature T_g. The wire has a diameter of 100 μm, a conductivity of 100 W/m·K, and a volumetric heat capacity of 2.57×10^6 J/m³K.

The frequency of the heating current, ω rad/s, can be chosen from a wide range of values. The bulb designer is worried that the surface temperature of the filament may oscillate with the heating current if he chooses the wrong frequency: this could cause a visible flicker in the bulb. In this problem, you are asked to help him figure out what's likely to happen.

Make the following approximations:

- The convection and radiation losses at the surface of the wire can be represented by a constant, known, effective heat transfer coefficient, $h_{eff} = 750$ W/m²·K, to surroundings at T_g.
- The filament can be treated as a long straight wire, without coils or end losses.
- The electric current produces a uniform, but time varying, volumetric heating (W/m³) of the wire:

$$\dot{q}_v(t) = q_0 \sin^2 \omega t = \frac{q_0}{2}(1 - \cos 2\omega t)$$

Answer the following questions.

a. What determines whether there is a significant temperature difference between the wire's center and its surface? Write down a dimensionless group that describes this condition. Do you expect a significant temperature difference here?

b. What independent time scales apply to this situation? What is the significance of the ratio(s) of these time scales?

c. What frequencies are "low" frequencies from a thermal viewpoint? Write down an accurate equation for the surface temperature of the wire, $T_s(t)$, if the frequency is low.

d. What frequencies are "high" frequencies from a thermal viewpoint? Write down an accurate equation for the surface temperature of the wire, $T_s(t)$, if the frequency is high.

e. Find an equation for the wire temperature that is valid for all frequencies.

f. Propose, with justification, an appropriate frequency for the current.

3.9 An automobile's drum brake is shown in Figure 3.20. The brake consists of a cast iron drum with brake shoes inside. (The shoes are made from a proprietary composite material.) When the brake is applied, the stationary shoes are pressed against the rotating drum with a pressure p creating frictional drag. This friction causes heat to be dissipated at the interface of the shoe and drum at a rate of $q = (\mu p \omega D/2)$ W/m^2, where μ is the coefficient of sliding friction and ω is the angular velocity.

For this problem, take both the shoes and the drum to be 1.3 cm thick, neglect curvature, and assume them each to be one-dimensional. Let $T_\infty = 20°$C and $h_1 = h_2 = 100$ W/m^{2}K . Take the shoes to have a conductivity $k_{shoe} \approx$ 1.5 W/m·K and a thermal diffusivity $\alpha_{shoe} \approx 4 \times 10^{-7}$ m^2/s.

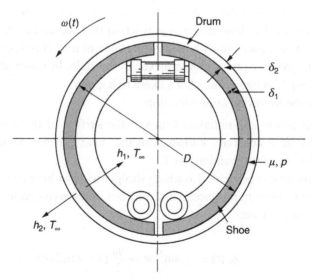

Figure 3.20. Brake drum and shoes (figure follows V.S. Arpaci [3.12]).

Answer the following questions, assuming the q is constant during the braking process, and that the shoe and drum are initially isothermal at T_∞. You may set $q = 3 \times 10^4$ W/m^2.

a. If the iron can be treated as a lumped object, derive an equation giving an upper bound on the temperature of the iron as a function of time, $T_{iron}(t)$.

b. For how many seconds must the brake be applied for the iron to reach a steady temperature under the conditions of the preceding question?

c. Now consider the initial stage of braking, immediately after the shoes contact the drum, during which semi-infinite behavior would be expected. Determine the time variation of the temperature at the interface of the shoes and the drum, $T_{interface}$.

d. Estimate the fraction of q that is transferred to the shoes, as opposed to the drum, during the initial period, and estimate the length of time for which this initial period persists.

3.10 A food processing company wishes to freeze blocks of spinach. A 30 cm thick slab of spinach is initially at 20°C. The slab of spinach is confined between a pair of flat plates outside of which flows a liquid coolant at −90°C [3.13, Problem 4-38]. The thickness or thermal conductivity of these plates can be chosen so as to create whatever thermal resistance, R_t (K · m^2/W), is desired. For this problem, you may assume the liquid heat transfer coefficient to be very large.

a. What is the minimum time required to bring the average temperature of the spinach, $\overline{T}$, to −30°C? What are R_t and the final surface temperature of the spinach in this case?

b. It turns out that the spinach must not go below −60°C at any point to prevent damage. One way to ensure this is to choose the thermal resistance needed so that the temperature gradients inside the spinach remain very small throughout the freezing process. How long does the spinach take to cool to $\overline{T} = -30$°C in the presence of this resistance?

c. By allowing temperature gradients in the spinach, we can make it cool faster than in Part **b**. Estimate the minimum thermal resistance that can be used *without* causing damage. Also find the time to cool in this case.

Estimated properties of spinach

thermal conductivity	$k = 0.5$ W/m·K
specific heat capacity	$c_p = 3000$ J/kg·K
density	$\rho = 900$ kg/m^3

3.11 In preparation for installing a green roof on a flat roof building, 5 cm of compacted soil is put over existing roof insulation. The next step is to begin planting. At the end of a summer night, the soil is at a uniform temperature of 15°C. At dawn the air temperature quickly increases to 25°C. A modest wind is blowing giving a surface heat transfer coefficient of 20 W/m^{2}K . The gardener needs a soil surface temperature of 20°C to start planting. The gardener would like you to make a first estimate of the soil surface temperature one hour after the air temperature rises to 25°C. In some instances, the sky will be clear with solar energy of 200 W/m² absorbed by the soil surface. In all cases neglect infrared radiation and evaporation from the surface.

a. Make an estimate of the lowest possible temperature of the soil surface temperature after 1 hour for two cases: with and without the solar energy.

b. Make an estimate the net energy transfer to the 5 cm thick soil layer after an hour. Does this estimate represent an upper or lower bound on the soil surface temperature?

3.12 When the space shuttle reenters the Earth's atmosphere, the high air speeds lead to a very high external gas temperature and heat transfer coefficient at the shuttle's outside surface. Portions of the shuttle have ceramic fiber insulation tiles on the exterior. Estimate the thermal time constant of the insulation and of the metal surface underneath. Use this to set up a simplified model for the temperature rise of the insulation tile and the metal skin during reentry heating, which can be assumed to last less than 1000 s. Assume a constant external gas temperature. Take the tile thickness as 1 cm, density of 144 kg/m³, specific heat of 630 J/kg·K, and effective conductivity of 0.08 W/m·K. Assume the metal is 1 cm thick and is aluminum.

3.13 During the manufacturing of 2 cm thick window glass, a glass sheet at initially at a uniform temperature of 400 °C is rapidly cooled on both sides by an array of air jets at 100°C. The jets produce a convective heat transfer coefficient of 120 W/m²K that is uniform over both surfaces of the glass. Make a quick estimate of the surface time-temperature history of the glass and the rate of heat transfer from the glass to the air. Radiation from the glass surface can be neglected for this calculation.

a. Your supervisor recommends that you used a lumped approximation for the transient response of the glass sheet. Will this approximation give good estimates for the surface temperature and heat flux?

b. Using the lumped approximation, will the estimated glass surface temperature at a given time be higher or lower than the correct value?

c. Will the estimated heat transfer rate be an upper or lower bound on the actual value?

3.14 A proposed system for measuring the thermal conductivity of an insulation sample is illustrated in Figure 3.21. A thin platinum foil (0.2 mm) is attached to the surface of a block of insulation that is 20 mm thick. Initially, the foil and insulation are at 20°C. A DC current is applied to the foil creating ohmic heating equivalent to 100 W/m². The platinum foil's resistance has been calibrated as a function of temperature, so that the foil temperature can be recorded during the test. The height and width of the insulation face in contact with the platinum is much larger than the insulation thickness, and the entire face is covered with platinum. Contact resistance between the platinum foil and the insulation can be represented by a contact conductance of $h_{contact} = 50$ W/m²K. Heat loss from the back side of the foil is negligible.

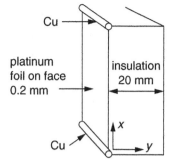

Figure 3.21. Conductivity measuring system.

Properties of the materials are: $(\rho c_p)_{\text{platinum}} = 2.6 \times 10^6$ J/m³K, $k_{\text{platinum}} = 71$ W/m·K, $(\rho c_p)_{\text{insulation}} = 3 \times 10^5$ J/m³K, and $k_{\text{insulation}} = 0.02$ W/m·K.

a. The experiment is to be run until the unheated side of the insulation just starts to experience a temperature change. Estimate this time.

b. Find an approximate expression for the foil temperature as a function of time and plot it. Indicate the foil temperature after 100 seconds of heating on the plot. Show the electrical analogy for heat transfer with appropriate thermal resistances. Clearly state your assumptions.

c. Suppose that you know the density and specific heat capacity of the insulation, but you do not know the contact resistance. How would you use the experimental data to determine the insulation's conductivity?

d. The top and bottom edges of the platinum foil are connected to large copper leads to deliver the current. The copper is much thicker than the platinum so that its temperature remains at the initial temperature 20°C during the entire experiment. Now we would like to predict the platinum temperature distribution in the *x* direction at each specific time in the experiment to allow for end effects. Make an approximate model with suitable assumptions that allow you to set up (but not solve) a simple differential equation for this temperature distribution. What are the boundary conditions? Write an expression for each term in the equation using the properties and geometry given in the problem.

3.15 Quenching, or reflooding, occurs when a liquid cools a very hot metal surface through a boiling process. A surface that is initially unwetted and at a temperature far above the liquid's saturation temperature, T_{sat}, is cooled by a falling film of liquid, as shown in Figure 3.22. The liquid film boils at its advancing edge causing liquid to sputter off the surface; but behind the quenching front, the liquid film is continuous and there is no phase change.

The location of sputtering moves downward at a speed u. The surface temperature at the location of sputtering, T_s, may be assumed to be known. For water at 1 atm, T_s is about 100 K higher than T_{sat}. Far ahead of the liquid, the dry surface has temperature T_w. The vapor phase is at T_{sat}. The heat transfer coefficient from the metal surface to the vapor, h, clearly varies along the surface, taking different values in the dry region, the short sputtering region, and the film region.

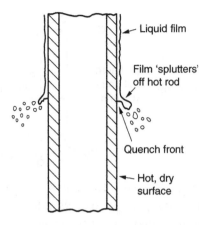

Liquid film

Film 'splutters'
off hot rod

Figure 3.22. Reflooding (figure from
Whalley [3.14]).

Quench front

Hot, dry
surface

In this problem, we consider the metal surface to be thin and to be adiabatic on the inner side (away from the vapor and liquid). The liquid film is assumed to have a uniform thickness δ, and the wall has a thickness t. The speed u is assumed to be constant. The situation is shown schematically in Figure 3.23.

a. Suppose that the metal wall is made of copper with thickness $t = 1$ mm, and the liquid film is water with $\delta = 1$ mm. What is the surface temperature of the film? What is the effective heat transfer coefficient (in W/m²K) between the metal surface and the bulk vapor in the region where the liquid film covers the wall? Will conduction within the metal in the y-direction be important?

b. The coordinate system shown in Figure 3.23 moves with the advancing liquid front, so that $x = 0$ is always the location of sputtering. Using these coordinates, derive a differential equation that describes $T(x)$ for the metal surface. (It's not necessary to solve the equation.)

c. Assuming that all the energy removed from the wall is taken up by latent heat, find a relationship between the velocity u of the quenching front and the bulk velocity of the falling film, u_b. Under what conditions is $u \ll u_b$? (It's not necessary to plug numbers into your formula.)

d. Suppose that the heat transfer coefficient has a value h_l under the liquid $(x < 0)$ and a value $h_v = 0$ under the vapor $(x > 0)$. Suppose further that

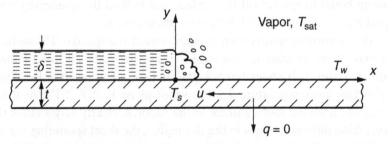

Figure 3.23. Configuration for reflooding. Coordinates move with the quenching front.

the quenching front moves slowly, in the sense that $\rho_w c_w t u / k_w \ll 1$, for ρ_w, c_w, and k_w the wall density, specific heat capacity, and thermal conductivity, respectively. Obtain a relationship between the wall temperature, T_w, and the quench front speed, u. Explain your model clearly. (It's not necessary to plug numbers into your formula.)

e. The two-region model of the preceding part was proposed by Yamanouchi [3.15] to explain his measurements. For the conditions described in Part (a) above, he had to take $h_l \approx 10^6$ W/m^{2}K to fit his data. Is this reasonable in view of your answer to Part (a)? Explain what the Yamanouchi model neglects, and describe how you might change it to get better results [3.16].

Modeling Convection

Convection is heat conduction into a medium that moves. The motion is generally nonuniform, and we are often most concerned with heat flow through a wall perpendicular to the direction of the flow.

The modeling of convection heat transfer always involves two elements. The first is to determine the local rate of heat transfer from wall to fluid. The second is to determine the streamwise energy balance on the fluid. Usually, these two elements are coupled because streamwise temperature variations affect the local temperature differences that drive heat transfer. Only occasionally will the rate of heat transfer be set independently, such as by electrical heating of the wall.

The determination of convection heat transfer rates always begins with the identification of the flow field, and, in particular, it requires one to distinguish between laminar and turbulent flow. Laminar flows often admit relatively simple modeling, in which an estimate of the local boundary layer thickness may be used to determine local convection thermal resistance by analogy to heat conduction. In turbulent flow, thermal resistance has a less obvious relationship to boundary layer thickness, and empirical data must usually be applied. The local laminar thermal resistance is relatively sensitive to the upstream thermal history and wall boundary conditions, whereas the turbulent thermal resistance is not.

For internal flows, bulk flow energy conservation provides a powerful tool for assessing streamwise variations in temperature and other properties. For external flows, the streamwise growth of the boundary layer is of primary importance, requiring consideration of the changing velocity field as well as the temperature history. Such problems can sometimes be approached using the integral energy conservation equation.

In addition to situations that have a mean flow past a wall, some situations of technical importance involve stirred tanks or free surfaces above turbulent pools in which heat transfer to the bulk is driven by turbulent mixing with negligible mean flow. Surface renewal models can be applied to estimate heat transfer coefficients in these cases if certain information about the mixing process is available.

Buoyancy-driven flows add a further level of complexity, in that the fluid velocity depends entirely on the temperature differences that are present during

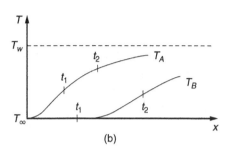

Figure 4.1. Two fluid particles traveling past a warm plate. (a) Particle trajectories showing positions at times t_1 and t_2. (b) Particle temperature histories.

convection. Here the velocity and temperature fields must usually be analyzed simultaneously.

Convective mass transfer, such as occurs in evaporation or dehumidification, has a strong similarity to convective heat transfer. The analogy between these two processes provides a tool for evaluating mass transfer rates.

In this chapter, we consider these issues, starting with basic ideas about convection thermal resistances.

4.1 Basic Concepts of Convective Heat Transfer Coefficients

We may think of convective heat transfer either in terms of the steady temperature distribution in a moving fluid or in terms of transient conduction into moving particles of fluid. Both approaches are useful in understanding how convective heat transfer coefficients depend on fluid properties, the velocity distribution, and upstream temperature variations. Let us begin by thinking about laminar boundary layers.

4.1.1 Convection in a Laminar Boundary Layer

Consider laminar flow of a cold fluid past a warm flat plate. For now, let the Prandtl number be unity, so that the thermal and momentum boundary layers have the same thickness. Two fluid particles, A and B, flow past the plate, with particle B being somewhat farther away from the surface (Fig. 4.1a). Particle A will travel more slowly than B because it is closer to the wall, where the velocities are lower. Particle A is also closer to the warm wall. As a result, on reaching any point x farther downstream, particle A will be hotter than particle B (Fig. 4.1b). Particle A will have taken longer to reach x as well. The rate at which the particle temperatures rise will clearly depend on their heat capacitance, ρc_p, their velocity, and the thermal conductivity, k, of the fluid, in addition to their location.

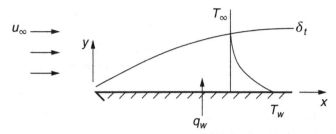

Figure 4.2. Temperature distribution in a laminar boundary layer.

The resulting temperature distribution in the y direction is shown in Figure 4.2. The fluid temperature drops smoothly from T_w to T_∞ across the thickness, δ_t, of the boundary layer. At the wall, the heat flux is the product of the local temperature gradient normal to the wall and the fluid conductivity. We may think of the heat flux as passing from T_w to T_∞ by conduction through the boundary layer. This is not strictly correct, of course, because the heat is being absorbed by the fluid particles moving past the wall in the boundary layer, but it provides a useful model of the process, much like the unsteady conduction resistance introduced in Section 3.6.

To develop this idea, we may write an expression for the temperature distribution in the y-direction in the boundary layer:

$$T(y) - T_\infty = (T_w - T_\infty) f(y/\delta_t) = \Delta T \, f(y/\delta_t) \qquad (4.1)$$

for $f(y/\delta_t)$ a function that decreases smoothly from a value 1 at $y = 0$ to a value 0 at $y = \delta_t$, and $\Delta T = T_w - T_\infty$. The thermal boundary layer thickness, δ_t, increases with x. The heat flux from the wall may be calculated using Fourier's law and Eq. (4.1):

$$q_w = -k \left. \frac{\partial T}{\partial y} \right|_{y=0}$$

$$= -k\Delta T \left. \frac{\partial f}{\partial y} \right|_{y=0}$$

$$= \frac{k\Delta T}{\delta_t} [-f'(0)] \qquad (4.2)$$

The exact form of $f(y/\delta_t)$ can be determined only by solving the boundary layer equations, but it is often approximately cubic in laminar flows. The value of $-f'(0)$ is about 3/2 for a flat plate boundary layer [4.1, Section 6.5], and this value is typical of many convection problems.

To a rough approximation, therefore, we may estimate the heat flux in terms of the boundary layer thickness for laminar flow

$$q_w \approx \frac{k\Delta T}{\delta_t} \qquad (4.3)$$

Equation (4.3) provides a versatile means of modeling laminar convective heat transfer because we are often able to think about factors affecting or bounding the boundary layer thickness, even if we are not able to estimate it precisely. This equation

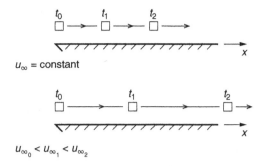

Figure 4.3. Effect of fluid velocity variation on particle trajectories showing location at times t_0, t_1, and t_2 for a flow at constant freestream speed and for an accelerating freestream.

gives us an expression for the convective heat transfer coefficient in a laminar flow, as well:

$$h \equiv \frac{q_w}{\Delta T} \approx \frac{k}{\delta_t} \qquad (4.4)$$

From this, we see that h is higher for thinner thermal boundary layers, lower for thicker thermal boundary layers, and increases with the fluid thermal conductivity. The latter immediately shows that heat transfer coefficients will be higher for liquids than for gases, other things being equal, since liquid conductivities are typically an order of magnitude greater than those for gases. The thermal boundary layer thickness itself may depend weakly on k, so that h may increase with k a bit more slowly than shown in Eq. (4.4). We shall return to Eqs. (4.3) and (4.4) often.

Changes to the flow that raise or lower the thickness, δ_t, of the warm layer near the wall will lower or raise the heat transfer coefficient. For example, if the flowing fluid has a larger ρc_p, it can absorb more heat with less temperature rise. This means that a fluid with a high ρc_p (a liquid, for example) will have a thinner thermal boundary layer than a fluid with a low ρc_p (a gas, for example).

Likewise, fluid particles moving at higher speed will have spent less time heating up when they reach a particular location than particles that have traveled there at lower speed. Raising fluid velocity therefore thins the thermal boundary layer and raises h. For the same reason, we see that a flow that accelerates in the streamwise direction will have locally higher values of h than one that decelerates from the same initial velocity (Fig. 4.3).

The average thickness of the boundary layer on a surface determines the average heat transfer coefficient, $\overline{h}$, on the surface. A long surface allows the boundary layer to grow thicker than a shorter surface, creating a lower $\overline{h}$. Thus, engineers sometimes replace a long fin in a heat exchanger with two shorter fins having the same total length. Each shorter fin has a higher average heat transfer coefficient than the longer fin because the shorter boundary layers remain thinner (Fig. 4.4).

The history of the wall temperature or heating also affects the heat transfer coefficient. We may compare the case of a constant wall temperature to that of a constant wall heat flux (Fig. 4.5). In the constant wall temperature case, a fluid particle A passing the wall will warm from T_∞ toward the wall temperature T_w as shown. In the

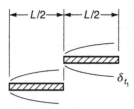

Figure 4.4. Comparison of long fin to two shorter fins of identical total length.

$\delta_{t_2} > \delta_{t_1}$

constant heat flux case, the wall itself becomes hotter in the streamwise direction, and a fluid particle B passing the wall will always lag behind the wall temperature and will not converge toward the wall temperature. The fluid in the constant heat flux case thus remains cooler relative to the wall than in the constant wall temperature case, resulting in a heat transfer coefficient that is higher. In fact, a comparison of the standard results for these two cases shows that h for the constant flux case is greater by 36%.[1]

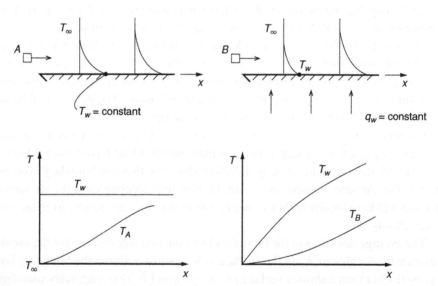

Figure 4.5. Comparison of fluid particle temperature for constant wall temperature and constant wall heat flux.

[1] For a laminar flow past a flat plate with Prandtl number Pr > 0.6, the results are: T_w constant, $\mathrm{Nu}_x = 0.332\,\mathrm{Re}_x^{1/2}\mathrm{Pr}^{1/3}$; q_w constant, $\mathrm{Nu}_x = 0.453\,\mathrm{Re}_x^{1/2}\mathrm{Pr}^{1/3}$. Procedures for calculating h when T_w varies arbitrarily are discussed in [4.2].

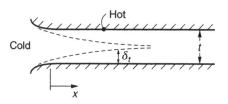

Figure 4.6. Boundary layer growth and heat transfer coefficient in the inlet of a channel.

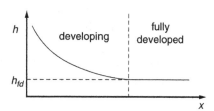

Example 4.1 *Laminar heat transfer coefficients for internal flow* We may use the preceding ideas to understand the streamwise variation of the heat transfer coefficient in an internal flow. Suppose that a cold fluid flows into a heated channel between two isothermal parallel plates a distance t apart (Fig. 4.6). At the inlet, a thermal boundary layer develops, and this boundary grows thicker as the fluid flows farther into the duct. At some distance downstream, the boundary layers from opposite walls merge at the centerline, and further boundary layer thickening is not possible. This begins the region of fully developed flow.

The usual practice for internal flow is to define h using the bulk temperature, T_{bulk}, in place of T_∞ in Eq. (4.1) so that ΔT is defined as $T_w - T_{bulk}$ in Eq. (4.4). Then, according to Eq. (4.4), the heat transfer coefficient in a channel will be high at the inlet, where the boundary layer is thin, and it will drop steadily as the boundary layer thickens. The boundary layers cannot grow past the centerline, however, so that the fully developed region has

$$\delta_t \approx t/2$$

From Eq. (4.4), it follows that the value of h in the fully developed region is

$$h_{fd} \approx \frac{2k}{t} = \text{constant}. \tag{4.5}$$

The exact solution for the fully developed heat transfer coefficient in laminar flow between parallel plates gives a constant value that is higher than the above by about a factor of 2, depending on the wall boundary condition. This difference is not surprising, given that we dropped the constant $[-f'(0)]$ in going from Eq. (4.2) to (4.3).

In fully developed flow, heat conducted from the wall is absorbed by fluid particles at different distances from the wall. The length $t/2$ in Eq. (4.5) represents the maximum distance of particles from the wall, and thus the equation gives a lower bound on h.

Note that the specific heat capacity does not affect the fully developed heat transfer coefficient. This may seem paradoxical, in so far as any heat transferred to the fluid *must* be absorbed by the fluid's heat capacitance, raising the bulk temperature of the fluid. The reason that ρ and c_p are not present in the expression for h is that no additional fluid is entrained in the thermal boundary layer in the fully developed region, so that the entire temperature profile of the boundary layer varies at the same rate when heat is added, without a change in its shape or thickness: neither δ_t nor $[-f'(0)]$ varies.

Example 4.2 *Laminar boundary layer on a flat plate* The exact result from boundary layer theory for heat transfer from an isothermal flat plate for Prandtl number $Pr > 0.6$ is

$$\mathrm{Nu}_x = \frac{hx}{k} = 0.332\,\mathrm{Re}_x^{1/2}\mathrm{Pr}^{1/3} \tag{4.6}$$

where $\mathrm{Re}_x = \rho u_\infty x/\mu$ is the Reynolds number, u_∞ is the constant freestream speed, μ is the dynamic viscosity, and $\mathrm{Pr} = \mu c_p/k$. We may examine this result to see how it reflects the trends we have discussed. If we write h in terms of dimensional variables, we have

$$h = 0.332\,k^{2/3}u_\infty^{1/2}x^{-1/2}(\rho c_p)^{1/3}(\mu/\rho)^{-1/6} \tag{4.7}$$

We see that h is greater for fluids having higher k (so that the heat flux is larger for a given temperature gradient near the wall), that it rises when u_∞ rises (so that fluid particles passing any given point have had less time to heat up), that it decreases as the boundary layer thickens in the x direction, that it is greater if ρc_p is greater (which keeps the thermal boundary layer thin), and that it weakly decreases as the kinematic viscosity μ/ρ increases (which thickens the momentum boundary layer).

We may also consider this result in terms of the boundary layer thickness. The 99% momentum boundary layer thickness is given by

$$\delta = \frac{4.92x}{\sqrt{\mathrm{Re}_x}}$$

and analysis shows that thermal boundary layer for this case is related to the momentum boundary layer as

$$\delta_t \cong \frac{\delta}{\mathrm{Pr}^{1/3}}$$

It follows from Eq. (4.6) that

$$h = 0.332\left(\frac{k}{x}\right)\mathrm{Re}_x^{1/2}\mathrm{Pr}^{1/3}$$

which we may rearrange as

$$h = (0.332)(4.92)k \left[\frac{\mathrm{Re}_x^{1/2}}{4.92x} \right] \mathrm{Pr}^{1/3}$$

$$= (0.332)(4.92)k \left(\frac{\mathrm{Pr}^{1/3}}{\delta} \right)$$

$$\cong 1.63 \left(\frac{k}{\delta_t} \right)$$

This is consistent with our previous results, Eqs. (4.2) to (4.4).

Example 4.3 *Heat transfer in slug flow vs. heat transfer in Poiseuille flow* The fully developed laminar velocity profile for a newtonian fluid in a circular tube is the well-known Poiseuille profile: a parabolic velocity distribution [4.1, Section 7.2]. For non-newtonian fluids, such as polymers or thick slurries, the fully developed velocity distribution may not be parabolic. Often a much blunter distribution is obtained, in which the velocity gradients near the wall are steeper and the flow in the center has a nearly flat profile. One model for this behavior is *slug flow*, in which the entire fluid moves at a uniform velocity, essentially sliding through the tube as a solid mass (think of mayonnaise) [4.3].

The Nusselt number for fully developed slug flow in an isothermal circular tube of diameter D is

$$\mathrm{Nu}_D = \frac{hD}{k} = 5.783 \tag{4.8}$$

(see Problem 4.1). In contrast, the Nusselt number for Poiseuille flow is about 45% lower:

$$\mathrm{Nu}_D = 3.657 \tag{4.9}$$

The principal reason for the higher heat transfer coefficient in slug flow is that the velocity near the wall is relatively higher than for Poiseuille flow. The Poiseuille flow has slow moving fluid near the wall which takes longer to reach a given location downstream than the faster moving near wall fluid of the slug flow. As a result, the near wall fluid of Poiseuille flow has come closer to the wall temperature. This fluid has less capacity to absorb additional heat, so heat must be conducted over a greater distance radially to reach the cooler fluid in the interior of the pipe. The thermal resistance is thus higher in Poiseuille flow than in slug flow, leading to a lower heat transfer coefficient. By a similar line of reasoning, a tube with a constant wall temperature will have a lower heat transfer coefficient than a tube with a constant heat flux wall condition.

4.1.2 Convection in a Turbulent Boundary Layer

For turbulent flow, the thickness of the fluid layer providing most of the convection thermal resistance is *much* smaller than the overall boundary thickness, owing to

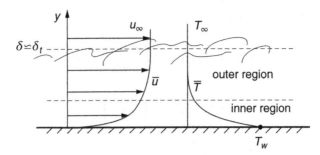

Figure 4.7. Temperature and velocity distributions within a turbulent boundary layer.

strong turbulent mixing within the boundary layer. This means that the we *cannot* estimate h using Eq. (4.4). Instead, empirical data are needed. In addition, the effect of streamwise variations on h is much weaker for turbulent flow than for laminar flow because turbulent mixing constantly changes the position of fluid particles with respect to the wall.

Figure 4.7 shows the structure of a turbulent boundary layer. The boundary layer may be considered to have two parts: an outer layer in which mixing by turbulent eddies is dominant and an inner layer in which the effect of the eddies dies out as the wall is approached and for which the flow comes to be dominated by fluid viscosity. The outer layer typically accounts for 80 to 90% of the total thickness of the boundary layer, and the inner layer accounts for 10 to 20% of it. Very near the wall, in the bottom 1% of the boundary layer, viscous effects are totally dominant and turbulent mixing is almost absent. This situation is the same for turbulent internal flows, for which we may think of the pipe radius as defining the boundary layer thickness [4.4].

In the outer region, strong turbulent mixing keeps the mean velocity and temperature close to the freestream values (or bulk values for a pipe flow). Here, the gradients are weak. In the inner region, turbulence is damped out and molecular conduction dominates heat transfer. The gradients become larger and reach their highest values in the viscous layer on the wall. Consequently, most of the thermal resistance and temperature drop in a turbulent boundary layer occurs very close to the wall, within the bottom few percent of the thickness. To obtain an upper bound for the heat transfer coefficient in turbulent flow, we might assume that the entire resistance is due to heat conduction in the viscous layer. That approach yields a local heat transfer coefficient that is typically between two and three times larger than measured values.

Consider a fluid particle traveling along a pipe in a turbulent flow. The particle is near the wall at some locations and near the core of the pipe in other locations. If it is heated when initially near the wall, then it will cool on reaching the core and be reheated when it again returns to the wall. As a result, the fluid particles near the wall at any particular location are likely to have lost any "history" of the upstream wall conditions because they have not remained near the wall for a long distance. This is very different from the situation for laminar flow illustrated in Figures 4.1 and 4.5. As a consequence, the local heat transfer coefficients in turbulent flow do not depend strongly on the upstream wall temperatures, and they are not strongly sensitive to the specific thermal boundary conditions. The heat transfer coefficient in turbulent pipe flow, for example, changes by less than 5% when the boundary condition is changed from one of uniform heat flux to one of uniform wall temperature [4.2, pp. 303–4, 319]. The only exception to this behavior is for very low Pr fluids, such as liquid metals, for which a high thermal conductivity weakens the thermal influence of turbulent mixing and increases the temperature gradient within the core of the duct.

Example 4.4 Turbulent boundary layer on a flat plate A correlation for local heat transfer through the turbulent boundary layer on a flat plate is[2]

$$\mathrm{Nu}_x = 0.032\,\mathrm{Re}_x^{0.8}\mathrm{Pr}^{0.43} \tag{4.10}$$

If we write h in terms of dimensional variables, we have

$$h = 0.032\,u_\infty^{0.8}k^{0.57}(\rho c_p)^{0.43}(\mu/\rho)^{-0.37}x^{-0.2} \tag{4.11}$$

Many of the trends are similar to those seen for a laminar boundary layer in Eq. (4.7), but there are notable differences as well: the dependence on x is significantly weaker, while the dependences on u_∞ and on (μ/ρ) are significantly stronger.

Example 4.5 Turbulent liquid flow in a circular tube A number of correlations have been developed for the heat transfer coefficient in fully developed turbulent pipe flow. The most accurate of these are due to Petukhov and Gnielinski (see Section 4.2). Gnielinski also developed simplified correlations applicable to limited ranges of Prandtl numbers [4.6] and having accuracies of 5 to 10% [4.7]. One of these is the following:

$$\mathrm{Nu}_D = 0.012\left(\mathrm{Re}_D^{0.87} - 280\right)\mathrm{Pr}^{0.4} \quad \text{for } 1.5 \le \mathrm{Pr} \le 500 \tag{4.12}$$

in which Re_D is the Reynolds number based on tube diameter D and bulk velocity u_b. If we assume that Re_D is large enough to neglect the subtracted term, we may again write h in terms of dimensional variables:

$$h \cong 0.012\,u_b^{0.87}k^{0.6}(\mu/\rho)^{-0.47}(\rho c_p)^{0.4}D^{-0.13} \quad \text{for liquids at high } \mathrm{Re}_D \tag{4.13}$$

The similarity of these dependences to those for a turbulent boundary layer [Eq. (4.11)] is striking. It is especially important to note that h is only weakly dependent upon tube diameter, in contrast to the behavior for laminar internal flow [cf. Eq. (4.5)]. This is because the thermal resistance is localized in the viscous region near the wall, rather than extending throughout the duct cross section as in laminar flow. For turbulent gas flow, the dependence of h on dimensional variables is very similar to that given by Eq. (4.13), but with a slightly lower exponent on u_∞ [see Eqs. (4.21)].

For fully developed laminar tube flow, according to Eqs. (4.4) and (4.5), the heat transfer coefficient is on the order of $k/(D/2)$, as the thermal boundary layer

---

[2] This equation applies for $2 \times 10^5 \le \mathrm{Re}_x \le 5 \times 10^6$ and $0.7 \le \mathrm{Pr} \le 380$ to about $\pm 15\%$ [4.5].

extends to the center of the pipe. To see what happens in turbulent flow, we may seek an effective thermal boundary layer thickness such that $h = k/\delta_{\text{eff}}$, where δ_{eff} represents the typical distance that heat flows by molecular conduction, so that

$$h = \text{Nu}_D \frac{k}{D} = \frac{k}{\delta_{\text{eff}}} \tag{4.14}$$

which yields

$$\delta_{\text{eff}} = \frac{D}{\text{Nu}_D} \tag{4.15}$$

The Nusselt number in turbulent flow will depend on both Reynolds and Prandtl numbers, but it is generally large, ranging from a minimum of about 10 (for Reynolds numbers near transition and Prandtl numbers of order 1) to more than 1000 at $\text{Re}_D = 10^6$ and $\text{Pr} = 0.6$. For higher Pr, the Nusselt number is even greater. It follows, then, that the effective thermal boundary layer thickness in turbulent pipe flow is a very small fraction of the pipe diameter. This fact is of considerable importance in modeling or controlling turbulent heat transfer.

4.2 Internal Flow Heat Transfer Coefficients

Heat transfer coefficients in tube flows have been studied extensively, for both laminar and turbulent flow. An enormous body of experimental and theoretical literature is available, as are detailed reviews of that literature and comprehensive correlations of data [4.8, 4.9, 4.10]. For fully developed flow in circular tubes with well-defined wall boundary conditions and constant physical properties, h can usually be predicted to within ±5 to 10% without difficulty. Often, however, the flow may not occur in a circular passage, may be developing, may have unusual boundary conditions, or may have physical properties that vary within the tube. In these latter cases, predictions of h from theory or correlations may be much less accurate, perhaps on the level of ±20 to 30% at best. (If the conditions are known in detail, computational fluid dynamics methods may provide better accuracy.)

In all internal flows, the heat transfer coefficient is defined with respect to the bulk temperature, T_b, and the wall temperature, T_w, as

$$h \equiv \frac{q_w}{(T_w - T_b)} \tag{4.16}$$

where q_w is the local heat flux going from the wall to the fluid.[3] The reason for using T_b, rather than a minimum or maximum temperature of the flowing fluid is that T_b represents the mixed-mean enthalpy of the flowing fluid and is therefore very easily incorporated into the bulk flow energy conservation equation (see Section 4.3 for discussion of the bulk flow properties and the bulk flow model).

---

[3] If viscous dissipation is significant, the wall temperature should be replaced with the *adiabatic wall temperature*, T_{aw}, in the definition of h, where T_{aw} is the temperature that an adiabatic tube wall would reach as a result of viscous dissipation (see Appendix A.4).

Table 4.1. *Laminar, fully developed Nusselt numbers based on hydraulic diameter [4.1].*

Cross section	T_w fixed	q_w fixed
Circular	3.657	4.364
Square	2.976	3.608
Rectangular		
$\quad a = 2b$	3.391	4.123
$\quad a = 4b$	4.439	5.331
$\quad a = 8b$	5.597	6.490
Parallel plates	7.541	8.235

4.2.1 Fully Developed Flow

Laminar internal flow. Typical Nusselt numbers for laminar fully developed internal flow are given in Table 4.1. For ducts of noncircular cross section, the Nusselt number, $\mathrm{Nu}_D = hD_h/k$, is based on *hydraulic diameter*, D_h,

$$D_h \equiv \frac{4A_c}{\mathcal{P}} \qquad (4.17)$$

where A_c is the cross-sectional area of the duct and $\mathcal{P}$ is the perimeter of the duct. Both the shape of the duct and the wall boundary condition have a significant effect on the heat transfer coefficient. In general, tube flow will be laminar if the Reynolds number based on hydraulic diameter and bulk velocity is less than about 2300.

As discussed in the previous section, the heat transfer coefficient for fully developed laminar flow depends on the thermal conductivity of the fluid and a dimension that characterizes the thermal boundary thickness, the latter being in general the hydraulic diameter:

$$h = \mathrm{Nu}_D \frac{k}{D_h} = \text{constant } \frac{k}{D_h} \qquad (4.18)$$

In fact, we see from the table that simply setting the constant in Eq. (4.18) equal to five will predict the laminar Nusselt number to within better than a factor of two for all cases. In the absence of more specific information, even such a crude estimate may be useful.

The literature contains laminar fully developed flow solutions for a wide range of boundary conditions, including convective heating or cooling, radiant heating or cooling, circumferentially varying heating, and cases involving viscous dissipation and/or axial conduction. An enormous (and perhaps absurd) variety of cross-sectional shapes have been analyzed as well [4.10].

Many engineering technologies involve flows through channels with diameters of centimeters or more. Very often, the flow rates in such systems are high enough that turbulent flow is the normal situation, and as a result the extensively literature on laminar flow has been viewed as somewhat academic. In recent years, however, microfrabrication technology has led to an increasing interest in flow through very small channels, and a renewed interest in laminar flow heat transfer.

Example 4.6 *Microchannel heat transfer coefficients* Microchannel heat transfer has significant applications in electronics thermal management and in some
high-performance heat exchangers. Laminar internal flow heat transfer coefficients in microchannels are substantially higher than for the centimeter-sized
passages encountered in conventional process equipment. Consider a circular
tube with constant wall heat flux, for which

$$h = \frac{4.364\,k}{D}$$

For air, with $k = 0.026$ W/m·K, a 2 cm diameter tube has $h = 5.7$ W/m^{2}K, whereas
a 100 μm microchannel has $h = 1.1$ kW/m^{2}K. For water, with $k = 0.60$ W/m·K, a
2 cm tube has $h = 130$ W/m^{2}K and a 100 μm microchannel has $h = 26$ kW/m^{2}K.
For liquid sodium, with $k = 80$ W/m^{2}K, the 100 μm microchannel would have $h =$
3.5 MW/m^{2}K! The latter case has been considered for use in high heat flux solar
energy systems. These microchannel heat transfer coefficients are well above the
ranges considered in Table 2.1, which refer to more conventional systems. Clearly,
such tables should not be used blindly.

The very large heat transfer coefficients of microchannels may lead to non-
negligible conduction resistance or fin effects in the walls of microscale convection systems. In some cases this means that the wall boundary condition will not
follow the simple designations of Table 4.1 [4.11].

Turbulent internal flow. Many correlations have been developed for turbulent internal flow, beginning in the early 1900s. These correlations were often based on power
laws that lent themselves to the computational tool of the times: the slide rule. The
slide rule is an analog, logarithmic calculator that was widely used by engineers and
scientists until the appearance of the pocket calculator in the early 1970s. It was an
ideal tool for multiplication, division, and exponentiation, but it was not useful for
addition and subtraction. The availability of this device made power-law correlations
convenient to use — whether they were theoretically appropriate or not!

The proper theoretical form of a correlating equation for heat transfer in turbulent internal flow had been identified by Prandtl in 1910, but it was not effectively
implemented until the late 1960s by Petukhov and coworkers [4.12]. Minor adjust-
ments by Gnielinski in 1976 led to what is now generally regarded as the standard
correlation for turbulent internal flow [4.6]:

$$\mathrm{Nu}_D = \frac{(f/8)(\mathrm{Re}_D - 1000)\mathrm{Pr}}{1 + 12.7\sqrt{f/8}(\mathrm{Pr}^{2/3} - 1)} \tag{4.19}$$

This equation predicts the best available experimental data for circular tubes to
within ±5 to 10% over the range $2300 \leq \mathrm{Re}_D \leq 5 \times 10^6$ and $0.5 \leq \mathrm{Pr} \leq 2000$ [4.7].
In this equation, f is the Darcy friction factor, given by

$$f = (1.82 \log_{10} \mathrm{Re}_D - 1.64)^{-2} \tag{4.20}$$

Both equations apply only to smooth-walled tubes.

As discussed previously, the precise wall boundary condition has only a weak effect on the turbulent flow heat transfer coefficient. Equation (4.19) may be used for either constant wall heat flux or constant wall temperature tubes, as well as for other similar situations in which the wall boundary condition varies in the axial direction.

For noncircular tubes, the hydraulic diameter may be used in place of D in Eqs. (4.19) and (4.20). The resulting prediction of h will be accurate to $\pm20\%$ in most cases. Better accuracy will be obtained for ducts that are more nearly circular and lack sharp corners. In cases such as very acute triangular cross sections, the prediction will be less accurate owing to the tendency for very sharp corners to damp turbulence locally. For some noncircular cross sections, specialized "effective diameters" have been developed which improve the accuracy of the prediction to within 5 to 10% [4.7].

For annular ducts in particular, complex schemes have been developed to account for differing boundary conditions on the inner and outer surface, and they are usually essential when the inner tube is much smaller in diameter than the outer tube. When the inner tube of an annulus is adiabatic, the hydraulic diameter can be used with the Gnielinski equation to predict h to within about $\pm13\%$ for any diameter ratio. For other cases, the hydraulic diameter approach should be used only when the inner diameter is within about 50% of the outer diameter if the annulus is concentric [4.7, Sections 4.7–4.8].

Gnielinski himself worked in the latter days of the slide-rule era, and he also proposed power-law approximations to his more general result (4.19). Those equations apply for limited ranges of Prandtl number, and are generally within 5 to 10% of Eq. (4.19)[4]:

$$\text{Nu}_D = 0.0214 \left(\text{Re}_D^{0.8} - 100\right) \text{Pr}^{0.4} \quad \text{for } 0.5 \le \text{Pr} \le 1.5 \tag{4.21a}$$

$$\text{Nu}_D = 0.012 \left(\text{Re}_D^{0.87} - 280\right) \text{Pr}^{0.4} \quad \text{for } 1.5 \le \text{Pr} \le 500 \tag{4.21b}$$

Note that the variation of Nu_D with Re_D is stronger for high Prandtl number fluids. For further discussion of power law approximations, see Appendix A.1.

4.2.2 Developing Flow and Axially Varying Boundary Conditions

Near the inlet of a duct, the flow is not fully developed and heat transfer coefficients are generally higher than for a fully developed flow (see Fig. 4.6). A number of correlations and models have been developed for the entry regions of duct flows, and most undergraduate textbooks include key results [4.1, Sections 7.2–7.3]. We will mention only the main points here.

For a laminar flow, as sketched in Figure 4.6, boundary layer models can be used to approximate the variation of $h(x)$ in the developing region, although exact (or numerical) solutions are also available for most cases of interest. The boundary layer solution for constant velocity flow over a flat plate does not strictly apply because

---

[4] Both fits have deviations of up to 25% for Reynolds numbers of 10^6 or more when $1 < Pr \le 2.5$.

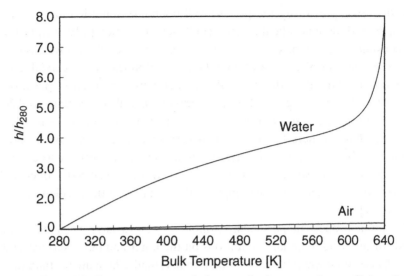

Figure 4.8. Effect of bulk temperature variation on the heat transfer coefficient for high Reynolds number turbulent flow of air and saturated water in a pipe at fixed mass flow rate and diameter; h is divided by its value for $T_{bulk} = 280$ K.

the developing momentum layer forces the main stream flow near the center of the channel to accelerate in order to conserve mass. Laminar entry lengths can be quite long, especially at high Prandtl number.

For turbulent flows, reference to correlations of experimental data is usually necessary. Turbulent entry lengths are generally short in terms of their effect on the local heat transfer coefficient, amounting to perhaps five diameters for high Pr and 15 to 40 diameters for Pr near one (depending on the inlet geometry). However, the effect of turbulent entry on the length-averaged heat transfer coefficient may persist to 100 diameters for Pr near 1.

In both laminar and turbulent developing flow, the bulk flow model describes the streamwise temperature variation (see Section 4.3). In cases of known wall heat flux, the heat transfer coefficient may not even be needed if only the bulk temperature change is of concern (i.e., if wall temperature need not be found).

When the wall temperature or heat flux varies in the streamwise direction, laminar heat transfer coefficients will not be the same as for uniform wall conditions. Extensive theory is available on the effect of varying wall temperature or flux on $h(x)$ [4.2]. For the turbulent case, reasonable accuracy may often be obtained by ignoring the effect on h of varying wall temperature or flux (i.e., by applying the fully developed h).

4.2.3 Flows with Varying Physical Properties

Temperature-dependent changes in fluid physical properties can have a substantial impact on the heat transfer coefficient, particularly for liquids. For example, simply changing the bulk temperature, with other parameters held constant, can affect the value of h. Figure 4.8 shows the influence of bulk temperature for high Reynolds number turbulent flow of air and water. The figure plots the heat transfer coefficient divided by its value at $T_{bulk} = 280$ K for bulk temperatures ranging up to 640 K. The mass flow rate and tube diameter are held constant, and Re_D is assumed to be large enough to neglect the subtracted terms in Eqs. (4.21). Physical properties are evaluated at the bulk temperature, and the wall temperature is assumed to be only slightly different from the bulk temperature.

For air, the effect of raising T_{bulk} amounts to only about 15% over this range of temperature. For water, in contrast, h increases by a factor of 8. Much of the increase in h occurs in the range above 600 K, as the critical point of water is approached (647.1 K); however, the heat transfer coefficient doubles when the bulk temperature increases from 280 K to 340 K, a range of just 60 K. The major factor influencing h for water is the nearly exponential decrease of dynamic viscosity μ with rising temperature; the other properties of water are not nearly as sensitive to temperature.

These results show both the importance of evaluating liquid properties at the correct temperature and the need to adjust the property values as the bulk temperature rises or falls along the length of the duct. For a first approximation, one may evaluate properties at the average bulk temperature along the duct.

Under some circumstances, physical properties may vary significantly between the wall temperature and the bulk temperature. In particular, this might occur at high heat flux, when the temperature difference is large. As before, liquids are more sensitive to such variations than are gases. A typical correction for this effect in liquids is to evaluate the Nusselt number at the bulk temperature and to multiply that value by a ratio of viscosity at bulk temperature to viscosity at wall temperature. For example, over the interval $0.025 \leq (\mu_b/\mu_w) \leq 12.5$, for turbulent liquid flow [4.1, Section 7.3]:

$$\mathrm{Nu}_D = \mathrm{Nu}_D\Big|_{T_b} \left(\frac{\mu_b}{\mu_w}\right)^n \quad \text{where } n = \begin{cases} 0.11 & \text{for } T_w > T_b \\ 0.25 & \text{for } T_w < T_b \end{cases} \tag{4.22}$$

Example 4.7 *Correlating heat transfer coefficients for supercritical tube flow*

Background. Liquids that are heated while at constant pressure below the thermodynamic critical point will undergo a phase change on reaching the saturation temperature. They boil and become gases, as shown in Figure 4.9 by the curve for p_{sub}. If the pressure at which heating occurs is above the thermodynamic critical-point pressure, however, no distinct phase transition occurs. A liquid heated at supercritical pressure will simply become less and less dense until it has the characteristics of a gas, as shown on the curve for p_{sup} in Figure 4.9. In the vicinity of the critical point, fluid properties vary strongly and rapidly with changing temperature. Property variations can consequently be very large in heat transfer systems operating at pressures slightly higher than the critical pressure.

Heat exchangers operate in the supercritical state in some steam power cycles and in some carbon dioxide refrigeration systems. Interest in CO_2 refrigeration systems has grown in recent years, particularly for automotive

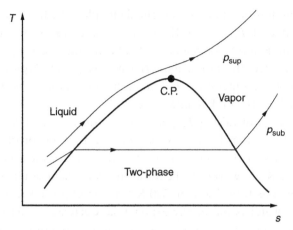

Figure 4.9. Heating a pure fluid from the liquid to the gas phase: p_{sub}, subcritical pressure; p_{sup}, supercritical pressure.

air conditioning systems. The basis of this interest is, somewhat paradoxically, that CO_2 refrigerants have a lower global warming potential than commonly used synthetic refrigerants such as 1,1,1,2-tetrafluoroethane (HCFC-134a), which was developed to replace refrigerants which had a high ozone depletion potential. Among the advantages of CO_2 are its low cost, low toxicity, abundant availability as an industrial by-product, and the generally smaller size of CO_2 compressors [4.13]. The disadvantages of CO_2 are a higher operating pressure (up to 100 atm versus 20 atm for HCFC-134a) and a lower theoretical Coefficient of Performance (COP) (2.8 vs. 5.5, although practical values may be closer to one another).

For automotive air conditioning systems, heat rejection will need to be possible for ambient temperatures exceeding 40°C, requiring that the working fluid in the heat exchanger be at a higher temperature. Since the critical-point temperature of CO_2 is 31.1°C, cooling must be done at supercritcal pressures (see Fig. 4.10). Heat exchangers of this type are sometimes called *gas coolers*.

Physical property variations with temperature at a constant and slightly supercritical pressure are shown qualitatively in Figure 4.11. As the temperature rises from the liquid-like state to the gas-like state, density, viscosity, and thermal conductivity drop substantially. The enthalpy rises, but it does so with an abruptly steeper slope at temperatures near the critical point. As a result, the specific heat capacity at constant pressure, $c_p \equiv (\partial h/\partial T)_p$, has a sharp spike. The temperature at the peak of this spike is called the pseudocritical temperature, T_{pc}. For water, the value of c_p at T_{pc} may be more than ten times that at a temperature just 10 K above or below T_{pc}.

In a tube flow, these property variations may either enhance or impair heat transfer. Consider, for example, a situation in which the tube wall temperature, T_w, is above T_{pc} while the bulk temperature, T_b, is below it. If T_w and T_b are not far apart, then much of the fluid within the passage will be a temperatures for which c_p is very high. This in turn allows a substantial heat absorption by the fluid with

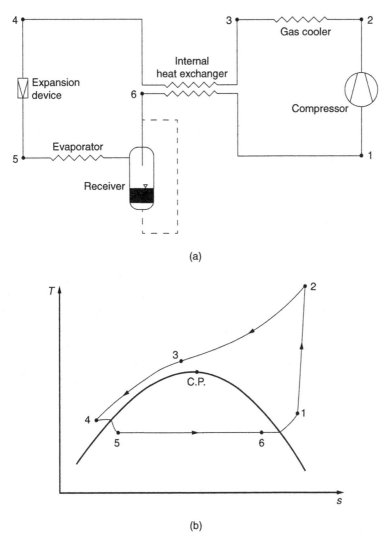

Figure 4.10. A carbon dioxide air conditioning cycle: (a) refrigeration circuit [4.14]; (b) cycle shown on a temperature-entropy diagram.

little resultant temperature increase. The heat transfer coefficient is consequently significantly greater than if c_p did not spike (see Fig. 4.12). On the other hand, if T_w is increased further, as by raising the wall heat flux, the temperature of the fluid in the tube cross section spans a greater range, and much less of it will be at the temperatures for which c_p spikes. Further, the fluid near the wall takes on gas-like properties, such as a low thermal conductivity and density. The result is that the heat transfer may be impaired at higher heat fluxes. Additional complexity may result from buoyancy effects associated with the strong density variation in the fluid, but in what follows we focus on the case when forced convection dominates the flow.[5]

---

[5] Jackson and Hall [4.15] suggest that a horizontal tube flow is free of buoyancy effects when $Gr_D/Re_D^2 < 0.001$, where $Gr_D = (g|\rho_w - \rho_b|D^3/v^2\rho_b)$ is the Grashof number based on tube diameter and the difference between wall and bulk density.

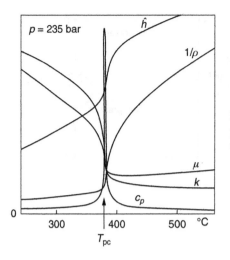

Figure 4.11. Temperature dependence of water's physical property variations at a constant supercritical pressure [4.15].

Correlation of Data. A substantial body of experimental data has been obtained for convection of water and carbon dioxide during turbulent flow in tubes at supercritical conditions. To correlate this data, the property variations just mentioned must be taken into account. One approach to this problem has been to adapt power law correlations with property-ratio corrections [4.15, 4.17], such as shown for normal liquids in Eq. (4.22). In the *absence* of variable property effects, power law correlations have the following form for Reynolds numbers above 10^4:

$$\mathrm{Nu}_D = a\,\mathrm{Re}_D^b\mathrm{Pr}^c \tag{4.23}$$

For the range of Prandtl number involved, Jackson and Hall [4.15] suggested that appropriate values of a, b, and c, were 0.0183, 0.82, and 0.5, respectively. (That

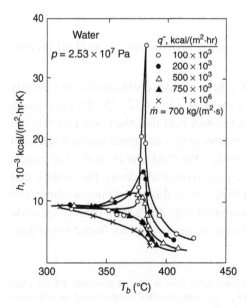

Figure 4.12. Enhanced and impaired heat transfer [4.16].

these values differ from those in Eqs. (4.21) simply reminds us of the limitations of power law fits.) The next step is to identify which property ratios should be used.

To see better the effect of specific physical properties, we may write this equation in terms of dimensional variables:

$$h = 0.0183\, u_b^{0.82} \rho^{0.82} c_p^{0.5} k^{0.5} \mu^{-0.32} D^{-0.17} \qquad (4.24)$$

For normal liquids, only the variation of μ is significant as temperature ranges between the wall and bulk values; but for supercritical liquids, all four physical properties may change greatly. The variation of k and μ with temperature is similar, however, and their exponents in Eq. (4.24) are close in magnitude and of opposite sign. Thus, the variations of k and μ can be ignored to a first approximation.

The density varies monotonically as temperature changes from its wall to its bulk value. Over a range of 20 K near the pseudocritical temperature, the density change of water can amount to a factor of 2. This effect is taken into account by introducing a density ratio correction factor to the Nusselt number, $(\rho_w/\rho_b)^d$, with d to be determined by fitting experimental data.

The specific heat capacity may also change dramatically, as noted before, but this variation need not be monotonic. Consequently, a simple correction factor $(c_{pw}/c_{pb})^e$ may not reflect the presence of the spike in c_p. Instead, an average value of c_p may be introduced in place of c_{pw}:

$$\overline{c_p} \equiv \frac{\hat{h}_w - \hat{h}_b}{T_w - T_b} = \frac{1}{(T_w - T_b)} \int_{T_b}^{T_w} c_p \, dT \qquad (4.25)$$

for $\hat{h}$ the specific enthalpy and with pressure taken to be uniform over the cross-section. This average will better indicate the mean c_p of the fluid over the cross-section of the pipe. The exponent e is again found by correlation of data.

Jackson and Hall, working from a set of approximately 2000 experimental observations, recommend the following final correlation:

$$\mathrm{Nu}_D = \mathrm{Nu}_D\big|_{T_b} \left(\frac{\rho_w}{\rho_b}\right)^d \left(\frac{\overline{c_p}}{c_{pb}}\right)^e$$

$$= 0.0183\, \mathrm{Re}_D^{0.82} \overline{\mathrm{Pr}}^{0.5} \left(\frac{\rho_w}{\rho_b}\right)^{0.3} \qquad (4.26)$$

for $\overline{\mathrm{Pr}} \equiv \mu_b \overline{c_p}/k_b$. This result correlates 78% of the water data and 79% of the CO_2 data to within $\pm20\%$; 91% of both sets of data correlated to within $\pm25\%$.

For further reading, see [4.15, 4.18].

4.3 Bulk Flow Energy Conservation Models

The one-dimensional bulk flow model is arguably the single most important model in the study of convection. In this section, we develop the steady, bulk flow energy

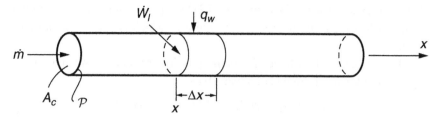

Figure 4.13. Steady flow through a duct.

conservation equation in several forms. Our objective is to find the temperature distributions in liquid and gas flows, and to understand the potential role of shaft work or viscous dissipation. For simplicity, we will neglect the effects of gravity and axial heat conduction.

Figure 4.13 shows a section of a duct through which a fluid flows steadily. The mass flow rate is $\dot{m}$ (kg/s), the duct cross-sectional area is A_c (m^2), the duct perimeter is $\mathcal{P}$ (m), and the coordinate x (m) is measured in the streamwise direction along the duct axis. A heat flux q_w (W/m^2) passes through the duct wall into the fluid; it may vary in the x-direction. Likewise, work transfer of $\dot{W}_l$ (W/m) may occur at a rate that varies with x, perhaps produced by an internal impeller or mixer. Both the cross section and the perimeter may also vary with x.

The x-component of fluid velocity, u, will usually have some distribution over the cross section of the duct, for example, a parabolic distribution in laminar flow or a much blunter distribution in turbulent flow. To avoid working with this distribution, we usually think in terms of a bulk or mass-average velocity, u_b, defined as follows:

$$u_b \equiv \frac{1}{\rho A_c} \int_{A_c} \rho u \, dA_c = \frac{\dot{m}}{\rho A_c} \tag{4.27}$$

Since $\dot{m}$ does not vary with x in a steady flow, this gives us a simple relationship with which to determine the bulk velocity: $\dot{m} = \rho u_b A_c$. In most situations, the density does not vary significantly over the cross section, so that averaging density across the area will not be necessary; exceptions include some supercritical fluid flows, as discussed in Section 4.2.3, and multiphase flows. These cases are beyond our scope.

4.3.1 Energy Conservation

We must consider energy in both its thermal and mechanical forms. We apply the first law to a section of the duct running from a location x to a location $x + \Delta x$. In a steady state, this mean that the total energy inflow rate equals the total energy outflow rate, accounting for the flow of enthalpy, $\hat{h}$ (J/kg), and kinetic energy, and for heat and work transfer. Enthalpy itself accounts for both internal energy and flow work of the fluid entering and exiting the control volume. For kinetic energy, we make the approximation that only the kinetic energy of the mean flow in the x-direction is significant; in particular, we will neglect the kinetic energy of turbulent eddies. The energy balance is as follows:

$$
\underbrace{\int_{A_c} \rho u \hat{h}\, dA_c \bigg|_x}_{\text{enthalpy entering}} + \underbrace{\int_{A_c} \rho u (u^2/2)\, dA_c \bigg|_x}_{\text{kinetic energy entering}} + \underbrace{q_w \mathcal{P} \Delta x}_{\text{heat transfer in}} + \underbrace{\dot{W}_l \Delta x}_{\text{work transfer in}}
$$

$$
= \underbrace{\int_{A_c} \rho u \hat{h}\, dA_c \bigg|_{x+\Delta x}}_{\text{enthalpy exiting}} + \underbrace{\int_{A_c} \rho u (u^2/2)\, dA_c \bigg|_{x+\Delta x}}_{\text{kinetic energy exiting}} \qquad (4.28)
$$

If we rearrange terms, divide through by Δx, and take the limit $\Delta x \longrightarrow 0$, the result is

$$
\frac{d}{dx} \int_{A_c} \rho u \hat{h}\, dA_c + \frac{d}{dx} \int_{A_c} \rho u (u^2/2)\, dA_c = q_w \mathcal{P} + \dot{W}_l \qquad (4.29)
$$

which shows that heat transfer and work transfer raise the flow rates of enthalpy and kinetic energy.

Like the velocity, the specific enthalpy usually has a distribution over the cross section of the duct, caused mainly by the variation of temperature over the cross section. To simplify Eq. (4.29), it is convenient to introduce the *bulk enthalpy*, $\hat{h}_b$, defined as the mass-average enthalpy of the flowing fluid:

$$
\dot{m}\hat{h}_b \equiv \int_{A_c} \rho u \hat{h}\, dA_c \qquad (4.30)
$$

Similarly, the kinetic energy integral in Eq. (4.29) defines the mass-average kinetic energy of the flowing fluid, $\overline{u^2}/2$:

$$
\dot{m}\frac{\overline{u^2}}{2} \equiv \int_{A_c} \rho u (u^2/2)\, dA_c \qquad (4.31)
$$

Equation (4.29) may now be written in final form as

$$
\boxed{\dot{m}\frac{d\hat{h}_b}{dx} + \frac{\dot{m}}{2}\frac{d}{dx}\overline{u^2} = q_w \mathcal{P} + \dot{W}_l} \qquad (4.32)
$$

This result applies to either laminar or turbulent flow.

Kinetic energy transport. The average kinetic energy is cumbersome, as it requires integration of the velocity distribution; however, with the density more-or-less uniform over the cross section, the integral may conveniently be calculated for particular cases, with the following results for circular ducts:

$$
\overline{u^2} = \begin{cases} u_b^2 & \text{for uniform flow} \\ 1.06 u_b^2 & \text{for fully developed turbulent flow} \\ 2 u_b^2 & \text{for fully developed laminar flow} \end{cases} \qquad (4.33)
$$

(For details, see Problem 4.9.) For turbulent flows, the approximation of a uniform velocity profile is usually sufficient.

Example 4.8 *One-dimensional compressible flow* Textbooks on compressible fluid flow usually treat duct flow as having uniform velocity because compressibility is most often important in gases moving at speeds comparable to the speed of sound. At such high velocities, turbulent flow is the most likely situation and the velocity profile is nearly uniform.

Assuming a uniform flow and neglecting shaft work, Eq. (4.32) becomes the usual equation for one-dimensional compressible flow with heat addition:

$$\dot{m}\frac{d}{dx}\left(\hat{h}_b + \frac{u_b^2}{2}\right) = q_w \mathcal{P} \tag{4.34}$$

If the heat addition rate is also 0, we obtain the usual equation for steady one-dimensional, adiabatic [nonisentropic] flow in a duct:

$$\frac{d}{dx}\left(\hat{h}_b + \frac{u_b^2}{2}\right) = 0 \tag{4.35}$$

Note that the flow under consideration may involve friction, causing kinetic energy to be dissipated into thermal energy, and it may include substantial pressure and density gradients in the flow direction, with reversible interconversion of mechanical and thermal energy. Yet the equation does not show separate terms for irreversible dissipation or reversible $p\,dv$ work. Because kinetic energy that is lost (or created) simply raises (or lowers) the thermal energy without affecting the total energy, the details of the interconversion process do not appear in the equation. Compressible flow in ducts is treated in detail in any number of textbooks (see, e.g., [4.19, Chapter 6]).

Bulk Temperature, T_b. For heat transfer calculations, we need to work with temperature, rather than enthalpy. The enthalpy may be eliminated in favor of temperature and pressure using a thermodynamic identity [4.20]

$$d\hat{h} = c_p\,dT + \frac{(1 - \beta T)}{\rho}\,dp \tag{4.36}$$

in which $\beta = -(1/\rho)(\partial\rho/\partial T)_p$ is the thermal expansion coefficient. This looks complicated, but for most purposes we just need to know that $\beta = 1/T$ for an ideal gas and that $\beta = 0$ for an incompressible substance.[6]

Equation (4.36) defines the *bulk temperature*, T_b, of the flowing fluid as the temperature that gives the bulk enthalpy at the local pressure from the equation of state: $h_b = h(T_b, p)$. In particular, if the pressure and c_p are uniform over the cross section, we may choose a convenient reference temperature T_{ref} and put $\hat{h} = c_p(T - T_{ref})$ in Eq. (4.30) to calculate T_b from the velocity and temperature profiles.

---

[6] The choice of T and p as independent variables is discussed further in Appendix A.2.

Before substituting Eq. (4.36) into the bulk flow energy equation, (4.32), we will identify the irreversible dissipation of mechanical energy so as to simplify the equations.

4.3.2 Dissipation of Mechanical Energy

The bulk flow energy equation incorporates the effect of shaft work in creating kinetic energy and in compressing the flow, both as reversible processes, but it includes as well the irreversible dissipation of mechanical energy. The latter causes temperature rise by viscous dissipation. Thermodynamic arguments (see Appendix A.3) show that the irreversible dissipation per unit length of duct is the shaft work minus the increase in kinetic energy minus the reversible flow work:

$$\dot{Q}_{\text{diss}} = \dot{W}_l - \frac{\dot{m}}{2}\frac{d}{dx}\overline{u^2} - \dot{V}\frac{dp}{dx} \tag{4.37}$$

where $\dot{V} = \dot{m}/\rho$. Substitution into the bulk flow energy equation Eq. (4.32) leads to

$$\dot{m}\frac{dh_b}{dx} - \dot{V}\frac{dp}{dx} = q_w\mathcal{P} + \dot{Q}_{\text{diss}} \tag{4.38}$$

Further, assuming that $\rho, p, c_p,$ and β are essentially uniform over the cross section of the tube, substitution of Eq. (4.36) into Eq. (4.38), gives:

$$\boxed{\dot{m}\left[c_p\frac{dT_b}{dx} - \frac{\beta T_b}{\rho}\frac{dp}{dx}\right] = q_w\mathcal{P} + \dot{Q}_{\text{diss}}} \tag{4.39}$$

In many cases, the dissipation rate can be calculated or shown to be negligible. Further, by introducing either the incompressible substance or ideal gas model, the second term can be simplified. These two steps lead to a greatly simplified equation.

Example 4.9 *Dissipation rate in fully developed pipe flow* An incompressible fluid flows steadily through a pipe of uniform cross section in the absence of work transfer. Under these conditions, the velocity profile does not change in the x-direction. Equation (4.37) becomes

$$\dot{Q}_{\text{diss}} = 0 + 0 - \dot{V}\frac{dp}{dx} \tag{4.40}$$

The pressure gradient is easily determined either from the friction factor (using Darcy's law) or from the balance of pressure force and shear force on a length of duct. In terms of the shear stress, τ_w, on the duct wall:

$$-\frac{dp}{dx} = \frac{\tau_w\mathcal{P}}{A_c} \tag{4.41}$$

For uniform density, $\dot{V} = u_bA_c$, so the dissipation rate per unit length of duct (in W/m) is

$$\dot{Q}_{\text{diss}} = u_b\tau_w\mathcal{P} \tag{4.42}$$

This quantity is the mechanical energy that is expended to overcome friction, that is, the pumping power per unit length of duct, and it serves to increase the temperature of the fluid.[7]

Example 4.10 *Dissipation in a developing laminar pipe flow* Now suppose that the flow enters a duct of length L with a uniform velocity profile and exits as a fully developed laminar flow. The inlet pressure is p_0 and the outlet pressure is p_L. The fluid along the centerline of the duct is accelerated by the pressure gradient in the developing section of the pipe, raising its kinetic energy. Integration of eqn. (4.37) with $\dot{W}_l = 0$ gives

$$\int_0^L \dot{Q}_{diss} \, dx = \dot{V}(p_0 - p_L) - \dot{m}\left(u_b^2 - u_b^2/2\right) \tag{4.43}$$

in which $\dot{V}$ is again constant, and kinetic energy is evaluated with Eq. (4.33). Rearranging, this result shows that some of the mechanical work done by pressure contributed to flow acceleration while the balance was dissipated by friction.

If the flow were fully developed at the entrance, the rightmost term would not appear. In a turbulent flow, Eq. (4.33) indicates that the acceleration effect would be only $\dot{m}(1.06\,u_b^2 - u_b^2)/2$. For sharp-edged ducts, the pressure loss in the entry region of may be dominated by dissipation, rather than acceleration, owing to the presence of flow separation at the inlet; for rounded entries, the opposite may be true.

Example 4.11 *Neglecting dissipation in fully developed laminar flow* To determine whether dissipation has a significant effect on bulk temperature, we may compare it to the rate of heat addition in Eq. (4.39). With Eq. (4.42),

$$\frac{\dot{Q}_{diss}}{q_w P} = \frac{\tau_w P u_b}{q_w P} = \frac{\tau_w u_b}{q_w} \tag{4.44}$$

If this ratio is small compared to unity, dissipation may safely be ignored. Otherwise, it must be included in Eq. (4.39). This statement is valid for turbulent flow as well.

If only the wall temperature is known (rather than q_w), we must introduce a heat transfer coefficient. For fully developed laminar internal flow, Eq. (4.18) shows that

$$h = C_1 \frac{k}{D_h} \tag{4.45}$$

for C_1 a constant on the order of 5. Similarly, the Darcy friction factor, f, is given by

$$f = \frac{C_2}{Re_{D_h}} \tag{4.46}$$

---

[7] In general, additional pressure losses due to valves, bends, entries, and so on will contribute to the overall pumping power. The total dissipation can be computed from the total pressure drop, which can be obtained hydraulic calculations using minor loss coefficients for the duct: $\int_0^L \dot{Q}_{diss} \, dx = \dot{V}\Delta p$.

for C_2 a constant ranging from about 50 to 100 depending on the passage shape.[8] With these results,

$$\frac{\dot{Q}_{\text{diss}}}{q_w \mathcal{P}} = \frac{\tau_w u_b}{q_w} \tag{4.47a}$$

$$= \frac{(f/4)(\rho u_b^2/2)u_b}{h \, \Delta T} = \frac{C_2 \, u_b^2 (\rho u_b D_h)}{8 C_1 \, \text{Re}_{D_h} k \, \Delta T} \tag{4.47b}$$

$$= \left(\frac{C_2}{8 C_1}\right) \frac{u_b^2 \mu}{k \, \Delta T} \tag{4.47c}$$

The factor $C_2/8C_1$ is a constant whose value is roughly 2. The other factor determines the strength of viscous dissipation locally: if it is small, dissipation may be ignored. Because the temperature difference in the denominator varies in x, we examine the overall importance of dissipation by considering a group based on the temperature difference at the inlet, which is called the *Brinkman number*, Br:

$$\text{Br} \equiv \frac{u_b^2 \mu}{k \, |T_w - T_{b\text{in}}|} \tag{4.48}$$

Dissipation will be negligible if $\text{Br} \ll 1$.

In the case of an adiabatic passage or a very long isothermal walled passage, this Brinkman number is not relevant — one should instead calculate the bulk temperature rise created by dissipation, as in Example 4.12.

4.3.3 Equations for the Streamwise Bulk Temperature Variation

Bulk temperature variation for an incompressible liquid. For an incompressible liquid, ρ is constant and $\beta = 0$. Equation (4.39) may be written as

$$\boxed{\dot{m} c_p \frac{dT_b}{dx} = q_w \mathcal{P} + \dot{Q}_{\text{diss}} \quad \text{for an incompressible liquid.}} \tag{4.49}$$

The dissipation term must be evaluated from Eq. (4.37). For fully developed flow in the absence of work transfer, Eq. (4.40) may be used for this purpose.[9]

Bulk temperature variation for an ideal gas. For an ideal gas, $\beta = 1/T$, and Eq. (4.39) has the form

$$\dot{m} \left[c_p \frac{dT_b}{dx} - \frac{1}{\rho_b} \frac{dp}{dx} \right] = q_w \mathcal{P} + \dot{Q}_{\text{diss}} \tag{4.50}$$

---

[8] For a circular tube, $C_2 = 64$; for parallel plates, $C_2 = 96$; and for a square duct, $C_2 = 57$ [4.10].
[9] The incompressible liquid model has been called into question by Kostić [4.21] and others [4.10, pg. 81]. For example, saturated liquid water has $\beta T = 0.083$ at 300 K, but at 500 K this has risen to $\beta T = 0.822$ – hardly zero! While the incompressible fluid model is reasonable for saturated water at 300 K, it is clearly not reasonable at 500 K. Check property data when in doubt.

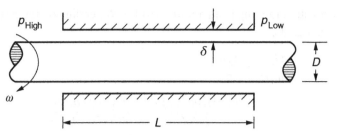

Figure 4.14. Oil flow in a journal bearing.

This equation shows that an ideal gas has coupling between thermal and mechanical effects, in the form of the pressure term. For a high-speed flow, this result is no more useful than Eq. (4.32) owing to the complications of evaluating the pressure gradient and dissipation terms (see Example 4.8). For low-speed flows, in which streamwise kinetic energy changes are negligible relative to streamwise enthalpy changes, we can obtain a simple equation by neglecting the kinetic energy term in Eq. (4.32) and putting $d\hat{h}_b = c_p dT$ for an ideal gas:

$$\dot{m}c_p \frac{dT_b}{dx} = q_w \mathcal{P} + \dot{W}_l \quad \text{for low speed ideal gas flow.} \tag{4.51}$$

In many cases, $\dot{W}_l$ is either a known external input or simply 0.

Example 4.12 *Dissipative heating in a journal bearing* A journal bearing consists of a stationary sleeve, a rotating shaft, and pressurized oil that acts as both a lubricant and a coolant (Fig. 4.14). Assuming that the shaft and its sleeve are adiabatic, determine the temperature of the oil leaving the bearing.

The details are as follows. The sleeve has a length $L = 5$ cm and the shaft has a diameter $D = 2$ cm. The clearance between the bearing and the rotating shaft is $\delta = 30$ μm. The gap between the sleeve and the shaft is filled with oil ($\rho = 860$ kg/m^3, $c_p = 2070$ J/kg·K, $k = 0.14$ W/m·K, $\mu = 0.0515$ kg/m·s). The shaft rotates at $\omega = 105$ rad/s, leading to a torque of $\tau = 0.0566$ N·m against the oil layer. The oil, which comes from a hydraulic reservoir, is pressurized at the left side of the bearing creating a pressure drop of $\Delta p = 33.7$ MPa between the left and right sides and causing oil to flow though the bearing at a rate of $\dot{m} = 1.6$ g/s. The oil enters at 320 K.

We will treat the oil as incompressible, so Eq. (4.49) applies, but with $q_w = 0$ for adiabatic walls:

$$\dot{m}c_p \frac{dT_b}{dx} = \dot{Q}_{\text{diss}} \tag{4.52}$$

An easy calculation shows that the entry lengths are very short, so that the flow is fully developed throughout essentially the entire bearing.

To determine $\dot{Q}_{\text{diss}}$, we use Eq. (4.37), integrating from $x = 0$ to $x = L$:

$$\int_0^L \dot{Q}_{\text{diss}} \, dx = \int_0^L \dot{W}_l \, dx - \dot{m} \frac{\overline{u^2}}{2} \bigg|_0^L + \dot{V} \, \Delta p \tag{4.53}$$

The first right-hand-side term is shaft power due to the moving shaft surface, $\omega\tau$. The second RHS term is the change in the kinetic energy of the oil, and, because the magnitude of the velocity vector is very low (on the order of 1 m/s), we shall neglect this change (see Problem 4.4). The third RHS term is positive because pressure decreases in the streamwise direction. Furthermore, $\dot{V} = \dot{m}/\rho$ for an incompressible fluid. Thus

$$\int_0^L \dot{Q}_{\text{diss}}\, dx = \omega\tau + \frac{\dot{m}}{\rho}\Delta p$$

$$= (105)(0.0566) + \frac{0.0016}{860}(33.7 \times 10^6) \tag{4.54}$$

$$= 5.943 \text{ N·m} + 62.70 \text{ N·m} = 68.64 \text{ J}$$

Since the flow is fully developed throughout the bearing, $\dot{Q}_{\text{diss}}$ will be constant in x and the integral in Eq. (4.54) is just $\dot{Q}_{\text{diss}}L$. Hence

$$\dot{Q}_{\text{diss}} = \frac{68.64 \text{ J}}{0.05 \text{ m}} = 1373 \text{ J/m} \tag{4.55}$$

The outlet bulk temperature is obtained by integrating Eq. (4.52):

$$\dot{m}c_p(T_{b\text{out}} - T_{b\text{in}}) = \int_0^L \dot{Q}_{\text{diss}}\, dx \tag{4.56}$$

so that

$$T_{b\text{out}} = T_{b\text{in}} + \frac{1}{\dot{m}c_p}\int_0^L \dot{Q}_{\text{diss}}\, dx$$

$$= 320 + \frac{68.64}{(0.0016)(2070)} = 340.7 \text{ K} \tag{4.57}$$

4.3.4 Solutions for the Bulk Temperature Variation

We may compute the bulk temperature variation of a flowing fluid using whichever of Eqs. (4.39), (4.49), or (4.50) is appropriate. In many cases, we can neglect viscous dissipation (Brinkman number $\ll 1$), and neglect the effects of streamwise pressure gradients on the state of compressible fluids. In those cases, all three of the mentioned equations reduce to

$$\dot{m}c_p\frac{dT_b}{dx} = q_w\mathcal{P} \tag{4.58}$$

This equation applies to gases or liquids under the stated conditions, and these conditions would render $\dot{W}_l$ negligible in Eq. (4.51).

For a known heat flux distribution, $q_w(x)$, the solution is

$$T_b(x) = T_{b\text{in}} + \int_0^L \frac{q_w(x)\mathcal{P}}{\dot{m}c_p}\, dx \tag{4.59a}$$

For a constant flux, this is simply

$$\boxed{T_b(x) = T_{b\text{in}} + \frac{q_w \mathcal{P}}{\dot{m}c_p} x}$$ (4.59b)

which is a straight-line variation of T_b.

For a known wall temperature distribution, we write $q_w(x) = h(x)[T_w(x) - T_b(x)]$. For laminar flow, $h(x)$ may be substantially affected by streamwise variations in the wall temperature; for a turbulent flow, that effect is much weaker (see Section 4.1). In either case, h will be higher in the developing region than in the fully developed region. Equation (4.58) becomes

$$\dot{m}c_p \frac{dT_b}{dx} = h(x)\mathcal{P}[T_w(x) - T_b(x)]$$ (4.60)

which has the solution

$$T_b(x) = T_{b\text{in}} e^{-\xi(x)} + e^{-\xi(x)} \int_0^L T_w(x) \left(\frac{h(x)\mathcal{P}}{\dot{m}c_p}\right) e^{\xi(x)} dx$$ (4.61a)

in which

$$\xi(x) = \int_0^x \frac{h(x)\mathcal{P}}{\dot{m}c_p} dx$$ (4.61b)

This, of course, looks just awful. For constant c_p and $\mathcal{P}$, however, $\xi(x)$ simply defines the average value of h from $x = 0$ to the current position (for example, accounting for entry effects):

$$\xi(x) = \frac{\mathcal{P}}{\dot{m}c_p} \int_0^x h(x)\, dx \equiv \frac{\bar{h}\mathcal{P}}{\dot{m}c_p} x$$ (4.61c)

In the case of constant T_w, the result simplifies greatly, to

$$\boxed{T_b(x) = T_w + \left(T_{b\text{in}} - T_w\right) \exp\left(-\frac{\bar{h}\mathcal{P}}{\dot{m}c_p} x\right)}$$ (4.61d)

which is an exponential decay from $T_{b\text{in}}$ to T_w.

Variation of the bulk temperature when dissipation is not negligible is discussed in Appendix A.4.

4.4 Integral Energy Conservation Models

The integral energy conservation model can be applied for boundary layer problems. In this model, the difference between the energy of fluid in the boundary layer and that of fluid outside the boundary layer is written in terms of an integral over the boundary layer thickness. Through energy conservation, the streamwise rate of change of this integral is related to the rate of heat addition at the wall. A polynomial model for the temperature distribution in the boundary layer is introduced, and, after some manipulation, a differential equation for the thermal boundary layer thickness is obtained. Once the boundary layer thickness is determined (as a function of position on the wall), the heat transfer coefficient and streamwise temperature distribution are easily found. The accuracy of the result is dependent on how closely the polynomial model for the temperature distribution agrees with the actual distribution.

Elementary examples of the integral energy conservation models may be found in Lienhard and Lienhard [4.1, Sections 6.5, 8.2]. More complex examples are given in the book by Kays et al. [4.2].

4.5 Surface Renewal Models

In a number of situations involving turbulent flow or rotating machinery, the flow fields are complex and do not resemble common turbulent boundary layer flows. Estimates for heat transfer coefficients at surfaces are still needed, and so modeling is essential. In some cases, the turbulent motion can be regarded as periodically scraping away fluid near a surface, which may have started to approach a boundary temperature, and replacing the near surface fluid with interior fluid at the bulk temperature of the turbulent medium. If the timescale of this surface renewal can estimated, a semi-infinite body heat conduction model may be adopted during the time between renewals, and by averaging the heat transfer during each period, an effective heat transfer coefficient may be obtained. These models are often credited to Higbie [4.22], although others have made similar estimates for related situations. We illustrate with an example.

Example 4.13 *Scraped surface heat transfer* In some mixing processes, a low-conductivity viscous material is held in a container having heated or cooled walls. The material is stirred by a rotating mixer within the container which periodically scrapes the heated or cooled material from the walls and replaces it with material at the interior bulk temperature, T_b. After the blade moves away, the viscous material at the wall is essentially stationary. These situations arise in processing of foods and polymers and may occur in either batch mixing containers or along the axis of screw extruders. The walls are typically some type of steel.

The heat transfer process at the wall restarts with each pass of the mixer's blade. Warm interior material is abruptly brought into contact with the cold metal wall, and heat diffuses into it essentially as if it were a semi-infinite body with step changed wall temperature. Because $\sqrt{k\rho c_p}$ for the metal wall is substantially greater than for the material inside, the wall's temperature changes little in response to the transient conduction process. We can treat the wall as having a constant temperature T_w. If our aim is to estimate the rate of heat loss to through the walls, we may focus our attention on the *time average* behavior of the process (Fig. 4.15).

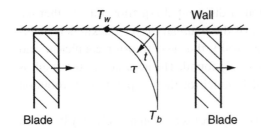

Figure 4.15. Temperature profiles at a scraped surface.

To do this, we calculate the average heat flux, $\bar{q}_w$ to the material inside, using the semi-infinite body result and assuming the time between successive passes of the blade is τ:

$$\bar{q}_w = \frac{1}{\tau} \int_0^\tau \frac{k(T_w - T_b)}{\sqrt{\pi \alpha t}} \, dt \tag{4.62}$$

$$= \frac{2k\Delta T}{\sqrt{\pi \alpha \tau}} \tag{4.63}$$

This result is probably more conveniently stated in terms of the frequency, $f = 1/\tau$, of the blade and the "thermal effusivity":

$$\bar{q} = \frac{2}{\sqrt{\pi}} \sqrt{k\rho c_p} \sqrt{f} \, \Delta T \tag{4.64}$$

For calculations that include the thermal resistance of the wall and an external cooling system, it is convenient to define an effective heat transfer coefficient for the scraped surface such that $\bar{q}_w = h_{eff} \Delta T$:

$$h_{eff} = \frac{2}{\sqrt{\pi}} \sqrt{k\rho c_p} \sqrt{f} \tag{4.65}$$

This physical model for extrusion is due to Jepson [4.23].

Scraped surface heat transfer coefficients may be quite large. For example, a polyethylene melt at 200°C has $k = 0.25$ W/m·K, $\rho = 770$ kg/m³, and $c_p = 2700$ J/kg·K. At a rotational frequency of 600 rpm (10 Hz), the effective heat transfer coefficient is $h_{eff} = 2570$ W/m²K .

It may be noted that the heat transfer coefficient obtained can be viewed as describing steady heat transfer through the time-mean flux diffusion thickness:

$$h_{eff} = \frac{k}{\delta_{mean}} \tag{4.66}$$

for δ_{mean} defined as

$$\delta_{mean} = \frac{1}{2}\sqrt{\pi \alpha \tau} \tag{4.67}$$

For the polyethylene case described above, $\delta_{mean} = 97$ μm.

Two additional factors can complicate the analysis of these devices.

1. Scraped surface equipment normally has a clearance between the tip of the moving blade and the wall. This clearance might be $s = 0.1$ to 1 mm, and the

material within the clearance is not scraped away. Consequently, an adhesive layer resides on the wall and creates additional thermal resistance. For example, referring again to the polyethylene processing system, a 0.1 mm adhesive layer has a conductance $k/s = 2500$ W/m²K — a thermal resistance as large as that of the periodically scraped layer.

2. The materials processed in these systems are often highly viscous and when the adhesive layer is sheared by the passing blade tip, considerable heat may be dissipated in the adhesive layer. Since the heat is generated within the thermal boundary layer at the wall, it will have a significant impact on the temperature profiles and heat transfer through that region. One way to account for this type of heating is to base the heat loss on an adiabatic wall temperature, similar to that used in high-speed boundary calculations.

4.6 Buoyancy-Driven Flows

We now turn our attention to buoyancy-driven flows where the fluid motion is driven by the density difference between fluid near a heated (or cooled) surface and fluid in the surrounding region. In these instances, fluid motion driven by external forces is absent.

Consider first laminar flow along a vertical heated plate, as shown in Figure 4.16. We expect temperature distributions somewhat similar to those shown in Figures 4.1 and 4.2, with heat transfer correlations analogous to those given in Eq. (4.6). However, there is not an obvious characteristic velocity to use in the Reynolds number. In addition, the velocity profile in the boundary layer will be different in buoyant flow than in forced flow. In the buoyant case, the velocity will go to 0 outside the thermal boundary layer as well as at the plate surface.

Suppose that the heated plate is at uniform temperature, T_p, and is surrounded by fluid at temperature T_∞ which is initially at rest. The temperature of the fluid close to the plate surface will be increased and the fluid density will decrease. A single fluid element, as shown in the figure, will experience forces in the vertical direction due to pressure, gravity, and viscous effects. Because the thermal boundary layer is thin, the vertical pressure gradient close to the plate will be the same as the vertical pressure gradient in the undisturbed fluid, given by hydrostatics as $\rho_\infty g$. To obtain an upper bound on the net vertical force acting on the fluid element, we will neglect viscous forces and set the element's temperature to that of the plate. Thus, the net force acting on a fluid element of volume $dA\,dx$ is

$$F = \left(\rho_\infty g - \rho_{T_p} g\right) dA\,dx \approx \beta \rho_{T_p} g(T_P - T_\infty) dA\,dx \qquad (4.68)$$

where β is the volumetric thermal expansion coefficient defined as

$$\beta = -\frac{1}{\rho}\frac{\partial \rho}{\partial T}\bigg|_p \qquad (4.69)$$

Now, assuming that the fluid element remains at the plate temperature as it moves vertically upward, the vertical force remains constant. If the element starts at rest at

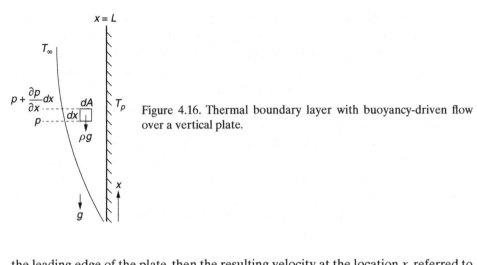

Figure 4.16. Thermal boundary layer with buoyancy-driven flow over a vertical plate.

the leading edge of the plate, then the resulting velocity at the location x, referred to as the buoyant velocity, can be found from elementary dynamics as

$$V_B = \sqrt{2g\beta(T_P - T_\infty)x} \qquad (4.70)$$

This represents an upper bound on the actual velocity set up within the boundary layer close to the plate. We can treat V_B, or more simply $V_B/\sqrt{2}$, as a characteristic velocity that can be used to form a dimensionless parameter similar to the Reynolds number. This parameter is known as the Grashof number, defined as

$$\mathrm{Gr}_x = \left[\frac{\rho x\sqrt{g\beta(T_P - T_\infty)x}}{\mu}\right]^2 = \frac{g\beta(T_P - T_\infty)x^3}{\nu^2} \qquad (4.71)$$

The Grashof number may be thought of as a buoyant flow equivalent to the square of the Reynolds number. For a laminar boundary layer, we would expect the heat transfer coefficient to have the same general form as we found for laminar forced convection on a flat plate. In particular, the Nusselt number should be a function of the Grashof number and the Prandtl number.

A large collection of experimental data for the average heat transfer coefficient over a vertical plate of length L has been closely correlated by Churchill and Chu [4.24] for $(\mathrm{Gr}_L\,\mathrm{Pr}) < 10^9$ and all values of Pr as

$$\mathrm{Nu}_L = 0.68 + 0.67\,\mathrm{Gr}_L^{1/4}\mathrm{Pr}^{1/4}\left[1 + \left(\frac{0.492}{\mathrm{Pr}}\right)^{9/16}\right]^{-4/9} \qquad (4.72)$$

One of the remarkable differences when compared with laminar forced convection results, Eq. (4.7), is the more modest influence of the plate length. Here the heat transfer coefficient is proportional to $L^{-1/4}$. Although the thermal boundary layer continues to grow thicker with increased plate length, the fluid velocity is also increasing with distance, offsetting the effect.

For typical plate dimensions near room temperature, natural convection and radiation heat transfer are about the same order of magnitude. At modest values of the Grashof number, the buoyant velocity is small and the thermal boundary layer thickness may be significant compared to the plate length, bringing into question the usual boundary layer assumptions.

When a heated plate is inclined facing downward, the same expressions hold, with *g* replaced by its component parallel to the plate surface. When it is inclined upward, a boundary layer flow may not exist. With a heated horizontal plate facing down and having unobstructed ends, the usual boundary layer model does not hold. Instead, the thermal boundary layer is thickest under the center of the plate and thins out near the edge. This sets up a horizontal pressure gradient forcing the fluid to flow toward the edges.

Numerous experimental and analytical studies of natural convection have been published for a variety of additional external geometries such as, for example, a heated horizontal cylinder or a heated sphere. A number of studies have also addressed natural convection inside an enclosed space. These range from small voids in cellular material to large architectural interiors. For small voids, there is a critical Grashof number below which viscous forces dominate over buoyancy forces and natural circulation is suppressed. In those instances, heat transfer through the fluid within the space is primarily by molecular conduction. A planar horizontal enclosure heated from below has stable conduction until the Rayleigh number, the product of Grashof and Prandtl numbers, exceeds 1708; above that value, cellular circulation occurs. Enclosures bounded by two vertical isothermal flat plates at different temperatures will have convective motion at all values of the Grashof number. However, the molecular conduction is not noticeably enhanced until the Rayleigh number exceeds 2700. For example, with two vertical window panes enclosing air at typical temperatures, enhancement is observed only when the separation exceeds 12 to 18 mm.

When both buoyancy and forced flow coexist, one simple measure of their relative importance is the ratio of the Grashof number for buoyancy to the square of the Reynolds number for the forced motion. This is essentially equivalent to the square of the ratio of the buoyant velocity to the free stream velocity.

4.7 Convective Mass Transfer

Mass is transferred to or from surfaces by convection in many important situations, especially those involving water entering or leaving airstreams. Cooling towers, humidifiers, and dehumidifers are just a few common examples. Our focus here will be on conditions for which mass transfer rates may be calculated by analogy to convective heat transfer rates. The analogies will be developed with the help of dimensional analysis.

Consider a mixture of several different chemical species. The mass fraction, or percent by mass, of one component *i* in the mixture is given by the ratio the density of *i*, ρ_i (in kilograms *i* per unit volume), to the density of the entire mixture, ρ:

$$m_i = \frac{\rho_i}{\rho} \tag{4.73}$$

The density of a mixture with N components is simply

$$\rho = \sum_{i=1}^{N} \rho_i \tag{4.74}$$

In a stationary medium where mass transfer occurs solely by diffusion, the rate of diffusion of species 1 in a mixture composed with a second species, 2, is given by Fick's law

$$\vec{J_1} = -\rho D_{12} \nabla m_1 \tag{4.75}$$

where $\vec{J_1}$ is the diffusive mass flux of species 1 (kg/m²s) and D_{12} is the diffusion coefficient or mass diffusivity. The mass flux is the mass flow rate per unit area; in one dimension, this would be $\dot{m}_1/A = J_1$. Fick's law is analogous to Fourier's law for conduction in a stationary medium:

$$\vec{q} = -k \nabla T \tag{4.76}$$

4.7.1 Mass transfer in a flowing medium

Consider mass transfer in the presence of a flowing medium, for example, the flow at a speed U_0 over a flat plate of length L, where the concentration of a species at the surface, $m_{1,w}$, is elevated compared to its concentration in the free stream, $m_{1,\infty}$. A good example is the evaporation from a still water surface to a dry air stream blowing across it. At low rates of mass transfer, the mass transfer will approach a limit where the mass transfer from the surface is solely by diffusion. Near this limit, the mass transfer rate can be characterized by a mass transfer coefficient g_1 defined as

$$J_1 = g_1 \Delta m_1 = g_1 (m_{1,w} - m_{1,\infty}) \tag{4.77}$$

The mass transfer rate can also be determined from the gradient of the concentration normal to the surface

$$J_1 = -\rho D_{12} \left(\frac{\partial m_1}{\partial n} \right)_w \tag{4.78}$$

These relationships can be compared to those for heat transfer from the surface

$$q = h \Delta T = h(T_w - T_\infty) = -k \left(\frac{\partial T}{\partial n} \right)_w \tag{4.79}$$

Similarly, the mass conservation equation for species 1 may be compared to the energy equation to obtain a prediction of the mass transfer coefficient from analogous convective heat transfer results. The mass balance in two dimensions includes mass transport by convective motion as well as mass transfer by diffusion in the flowing medium. In two dimensions at steady state and without chemical reactions

$$u_x \frac{\partial m_1}{\partial x} + u_y \frac{\partial m_1}{\partial y} = D_{12} \left(\frac{\partial^2 m_1}{\partial x^2} + \frac{\partial^2 m_1}{\partial y^2} \right) \tag{4.80}$$

The energy equation at steady state and without volumetric heat sources is

$$u_x \frac{\partial T}{\partial x} + u_y \frac{\partial T}{\partial y} = \alpha \left(\frac{\partial^2 T}{\partial x^2} + \frac{\partial^2 T}{\partial y^2} \right) \tag{4.81}$$

Nondimensionalizing the equations will help identify the analogy between the two. The velocities and coordinates may be nondimensionalized as follows:

$$\bar{u}_x = \frac{u_x}{U_0}, \quad \bar{u}_y = \frac{u_y}{U_0}, \quad \bar{x} = \frac{x}{L}, \quad \bar{y} = \frac{y}{L} \tag{4.82}$$

We define the nondimensional mass fraction and temperature so that they range from unity at the surface to zero in the main flow:

$$\bar{m}_i = \frac{m_1 - m_{1,\infty}}{m_{1,w} - m_{1,\infty}}, \quad \theta = \frac{T - T_\infty}{T_w - T_\infty} \tag{4.83}$$

Substituting these definitions into the mass conservation equation yields

$$\bar{u}_x \frac{\partial \bar{m}_1}{\partial \bar{x}} + \bar{u}_y \frac{\partial \bar{m}_1}{\partial \bar{y}} = \frac{D_{12}}{U_0 L} \left(\frac{\partial^2 \bar{m}_1}{\partial \bar{x}^2} + \frac{\partial^2 \bar{m}_1}{\partial \bar{y}^2} \right) \tag{4.84}$$

$$= \underbrace{\left(\frac{D_{12}}{v} \right)}_{= \frac{1}{Sc}} \underbrace{\left(\frac{v}{U_0 L} \right)}_{= \frac{1}{Re}} \left(\frac{\partial^2 \bar{m}_1}{\partial \bar{x}^2} + \frac{\partial^2 \bar{m}_1}{\partial \bar{y}^2} \right) \tag{4.85}$$

In the final expression, we identify the Reynolds number, Re, and the Schmidt number, Sc, which is defined as the ratio of kinematic viscosity, v, to diffusion coefficient. For the energy equation, the nondimensionalized form is

$$\bar{u}_x \frac{\partial \theta}{\partial \bar{x}} + \bar{u}_y \frac{\partial \theta}{\partial \bar{y}} = \frac{\alpha}{U_0 L} \left(\frac{\partial^2 \theta}{\partial \bar{x}^2} + \frac{\partial^2 \theta}{\partial \bar{y}^2} \right) \tag{4.86}$$

$$= \underbrace{\left(\frac{\alpha}{v} \right)}_{= \frac{1}{Pr}} \underbrace{\left(\frac{v}{U_0 L} \right)}_{= \frac{1}{Re}} \left(\frac{\partial^2 \theta}{\partial \bar{x}^2} + \frac{\partial^2 \theta}{\partial \bar{y}^2} \right) \tag{4.87}$$

The coefficients of the right-hand terms are the inverse of the product of the Reynolds number and the Prandtl number, Pr.

The equations for the dimensionless mass fraction and for the dimensionless temperature will be the same when the Reynolds numbers are the same (to maintain the same velocity distribution) and if the Prandtl number is equal to the Schmidt number. Further, if the boundary conditions applied to the equations are the same, the solutions will be the same:

$$\theta(\bar{x}, \bar{y}) = \bar{m}_1(\bar{x}, \bar{y}) \text{ for equal Re, Pr = Sc, and equivalent boundary conditions}$$
$$\tag{4.88}$$

For example, the boundary condition on the surface could be one of uniform surface temperature ($\theta = 1$) and uniform mass fraction ($\bar{m}_1 = 1$).

When the solutions are identical, the gradients of the two solutions at the surface will be the same:

$$\left(\frac{\partial \overline{m}_1}{\partial \overline{n}}\right)_w = \left(\frac{\partial \theta}{\partial \overline{n}}\right)_w \tag{4.89}$$

where $\overline{n} = n/L$ is the dimensionless coordinate normal to the surface. Combining Eqs. (4.77) and (4.78)

$$g = -\frac{\rho D_{12}}{m_{1,w} - m_{1,\infty}} \left(\frac{\partial m_1}{\partial n}\right)_w = -\frac{\rho D_{12}}{L}\left(\frac{\partial \overline{m}_1}{\partial \overline{n}}\right)_w \tag{4.90}$$

Similarly from Eq. (4.79)

$$h = -\frac{k}{T_w - T_\infty}\left(\frac{\partial T}{\partial n}\right)_w = -\frac{k}{L}\left(\frac{\partial \theta}{\partial \overline{n}}\right)_w \tag{4.91}$$

Using Eq. (4.89) with (4.90) and (4.91) it follows that

$$\underbrace{\frac{hL}{k}}_{=\text{Nu}} = \underbrace{\frac{gL}{\rho D_{12}}}_{=\text{Sh}} \tag{4.92}$$

The Nusselt number, Nu, is equal to the dimensionless mass transfer coefficient, known as the Sherwood number, Sh. For a given geometry and boundary condition, using the relationship of the Nusselt number as a function of Reynolds and Prandtl number, we can find the Sherwood number by substituting the Schmidt number for the Prandtl number and using the same Reynolds number.

For example, consider the case of fully developed turbulent flow in a circular tube. The approximate Gnielinski correlation for heat transfer is

$$\text{Nu}_D = 0.012 \left(\text{Re}_D^{0.87} - 280\right) \text{Pr}^{0.4} \quad \text{for } 1.5 \le \text{Pr} \le 500 \tag{4.21b}$$

The corresponding Sherwood number correlation would be

$$\text{Sh}_D = 0.012 \left(\text{Re}_D^{0.87} - 280\right) \text{Sc}^{0.4} \quad \text{for } 1.5 \le \text{Sc} \le 500 \tag{4.93}$$

Keep in mind that this analogy is applicable only when the mass transfer rate is low enough that the velocity field is not significantly affected by mass transfer, so that the mass transfer at the surface is dominated by diffusion. At higher mass transfer rates, the mass transfer coefficient found by analogy to convective heat transfer may be either higher or lower than the actual value because the mass transfer itself changes the velocity profile in the boundary layer: it will be higher in cases when mass flow is toward the wall, causing a thinner boundary layer; and it will be lower in cases when mass flow is away from the wall, causing a thicker boundary layer.

For further treatment of mass transfer, the reader is referred to [4.1, Chapter 11].

REFERENCES

[4.1] J. H. Lienhard IV and J. H. Lienhard V. *A Heat Transfer Textbook*, 4th ed. Mineola, NY: Dover Publications, 2011. http://ahtt.mit.edu

[4.2] W. M. Kays, M. E. Crawford, and B. Weigand. *Convective Heat and Mass Transfer*, 4th ed. New York: McGraw-Hill, 2005.

[4.3] J. P. Hartnett and Y. I. Cho. Nonnewtonian fluids. In W. M. Rohsenow, J. P. Hartnett, and Y. I. Cho (eds.), *Handbook of Heat Transfer*. McGraw-Hill, New York, 3rd edition, 1998, Chapter 10.

[4.4] S. B. Pope. *Turbulent Flows*. Cambridge, UK: Cambridge University Press, 2000.

[4.5] A. Žukauskas and A. Šlanciauskas. *Heat Transfer in Turbulent Fluid Flows*. Washington, DC: Hemisphere, 1987.

[4.6] V. Gnielinski. New equations for heat and mass transfer in turbulent pipe and channel flow. *International Journal of Chemical Engineering*, **16**(2), 359–368, 1976.

[4.7] R. K. Shah and M. S. Bhatti. Turbulent convective heat transfer in ducts. In S. Kakaç, R.K. Shah, and W. Aung (eds.), *Handbook of Single-Phase Convective Heat Transfer*. New York: Wiley-Interscience, 1987, Chapter 4.

[4.8] W. M. Rohsenow, J. P. Hartnett, and Y. I. Cho (eds.). *Handbook of Heat Transfer*, 3rd ed. New York: McGraw-Hill, 1998.

[4.9] S. Kakaç, R. K. Shah, and W. Aung (eds.). *Handbook of Single-Phase Convective Heat Transfer*. New York: Wiley-Interscience, 1987.

[4.10] R. K. Shah and A. L. London. *Laminar Flow Forced Convection in Ducts*. New York: Academic Press Inc., 1978 (Supplement 1 to the series *Advances in Heat Transfer*).

[4.11] A. E. Bergles, J. H. Lienhard V, G. E. Kendall, and P. Griffith. Boiling and evaporation in small diameter channels. *Heat Transfer Engineering*, **24**(1), 18–40, 2003.

[4.12] B. S. Petukhov. Heat transfer and friction in turbulent pipe flow with variable physical properties. In T. F. Irvine and J. P. Hartnett (eds.), *Advances in Heat Transfer*, **6**, 530–564. New York: Academic Press, 1970.

[4.13] G. Lorentzen and J. Pettersen. A new, efficient and environmentally benign system for car air-conditioning. *International Journal of Refrigeration*, **16**(1), 4–12, 1993.

[4.14] J. Wertenbach, J. Maue, and W. Volz. CO_2 refrigeration systems in automobile air-conditioning. In *International Conference on Ozone Protection Technologies*, Washington, DC, 21–23 Oct. 1996.

[4.15] J. D. Jackson and W. B. Hall. Forced convection heat transfer to fluids at supercritical pressure. In S. Kakaç, and D. B. Spalding (eds.), *Turbulent Forced Convection in Channels and Bundles*, Vol. 2. Washington, DC: Hemisphere, 1979.

[4.16] S. Kakaç. The effect of temperature-dependent fluid properties on convective heat transfer. In S. Kakaç, R. K. Shah, and W. Aung (eds.), *Handbook of Single-Phase Convective Heat Transfer*. New York: Wiley-Interscience, 1987, Chapter 18.

[4.17] E. A. Krashnoshchekov and V. S. Protopov. Experimental study of heat in carbon dioxide in the supercritical region at high temperature drops. *Teplofizika Vysokikh Temperatur*, **4**(3), 389–398, 1966.

[4.18] A. F. Polyakov. Heat transfer under supercritical pressures. In J. P. Hartnett, T. F. Irvine, and Y. I. Cho (eds.), *Advances in Heat Transfer*, **21**, 1–53. New York: Academic Press, Inc., 1991.

[4.19] M. A. Saad. *Compressible Fluid Flow*, 2nd ed. Englewood Cliffs, NJ: Prentice-Hall, 1993.

[4.20] A. Bejan. *Convective Heat and Mass Transfer*, 2nd ed. New York: John Wiley & Sons, 1995, p. 12.

[4.21] M. Kostic. Analysis of enthalpy approximation for compressed liquid water. *Proceedings of the 2004 ASME International Mechanical Engineering Congress and Exposition*, Anaheim. New York: ASME, Paper no. IMECE2004-59357, 2004.

[4.22] R. Higbie. The rate of absorption of a pure gas into a still liquid during short periods of exposure. *AIChE Journal*, **31**, 365–389, 1935.

[4.23] C. H. Jepson. Future extrusion studies. *Journal of Industrial Engineering and Chemistry*, **45**, 992–993, 1953.

[4.24] S. W. Churchill and H. H. S. Chu. "Correlating equations for laminar and turbulent free convection from a vertical plate." *International Journal of Heat and Mass Transfer*, **18**(11), 1323–1329, 1975.

[4.25] *Standards of Tubular Exchanger Manufacturer's Association*, 6th ed. New York: Tubular Exchanger Manufacturer's Association, 1978.

[4.26] H. Schlichting. *Boundary Layer Theory*, 7th ed. New York: McGraw Hill, 1979.

[4.27] M. V. Zagarola, A. E. Perry, and A. J. Smits. Log laws or power laws: The scaling in the overlap region. *Physics of Fluids*, **9**(7), 2094–2100, July 1997.

[4.28] P. A. Nelson and T. R. Galloway. Particle-to-fluid heat and mass transfer in dense systems of fine particles. *Chemical and Engineering Science*, **30**, 1–5, 1975.

PROBLEMS

4.1 Compute the Nusselt number for thermally fully developed slug flow in an isothermal circular tube. *Hint:* Use results from Section 3.3 to do this, explaining why that's appropriate and what it meant by thermally fully developed.

4.2 Water at 25.0 MPa flows through a 7.00 mm ID horizontal stainless steel tube at a mass flow rate of 0.100 kg/s. At a location where the bulk temperature is 375°C, the wall temperature is 385°C. Determine the heat flux through the tube wall, taking account of variable property effects. The wall roughness is approximately 1.5 μm. *Note:* Accurate property data should be used, e.g. from http://webbook.nist.gov/.

4.3 Helium gas flows through a packed bed of copper beads at a mass flow rate of 10 g/s. The beads are 1 mm in diameter and the porosity (open volume/total volume) of the bed is 40%. The helium is at 40.8 atm and 30°C as it enters the bed. The cross-sectional area of the bed is 2.5×10^{-4} m^2.

Estimate the average heat transfer coefficient between the helium gas and the beads. Do this by using both internal and external flow models. Then compare your result to a correlation for a packed bed.

4.4 Consider the bearing described in Example 4.12, in which the bearing surfaces were taken to be adiabatic. In practice, heat will be transferred from the oil to the bearing surfaces. In this problem, take the shaft and sleeve to be excellent heat conductors having a temperature of 320 K.

a. Estimate the kinetic energy change of the oil passing through the bearing, accounting for both the axial and tangential motions. Assume the oil enters the bearing at a uniform speed with no tangential velocity. How much does this energy change reduce the dissipation calculated in Eq. (4.54)?

b. Estimate the heat transfer coefficient h between the bearing surfaces and the oil.

c. Determine the bulk temperature of the oil leaving the bearing.

4.5 A proposed working fluid for a heat transfer device consists of a slurry of tiny paraffin spheres suspended in water. Each sphere will be enclosed in a very thin plastic capsule so that when the paraffin melts the spheres cannot coalesce. The melting temperature of the paraffin is T_{mp}, and the enthalpy of the mixture, $\hat{h}$, is sketched as a function of temperature in Figure 4.17. The density of the slurry is constant.

This fluid is expected to have some advantages if used in a temperature range that includes the melting temperature of the paraffin. To evaluate this, consider laminar flow of this fluid between two parallel plates. Assume that the velocity profile is everywhere fully developed with

$$u(y) = 6u_{av}\left[\frac{y}{b} - \left(\frac{y}{b}\right)^2\right]$$

for y the distance from the bottom wall and b the separation of the plates. Further, suppose that the fluid enters a heated section of the passage at $T_{in} < T_{mp}$ and that the walls of the heated section are held at $T_w > T_{mp}$.

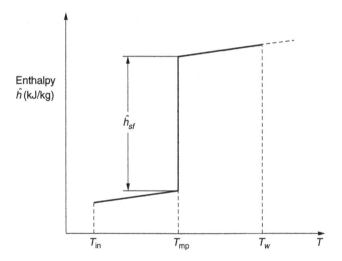

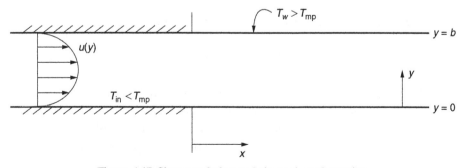

Figure 4.17. Slurry enthalpy and channel configuration.

a. Would you expect the heat transfer coefficient to be higher or lower than for an otherwise identical fluid that did not have a phase change? Explain briefly.

b. Suppose that $\delta(x)$ is the distance from the lower wall at which the fluid reaches T_{mp}. Sketch qualitatively the variation of $\delta(x)$ in the channel.

c. At some location where $\delta < b/2$, sketch the temperature profile in the fluid, $T(y)$. Give an approximation for the local wall heat flux in terms of δ and $(T_w - T_{mp})$.

d. Use an appropriately simplified version of Eq. (4.29) with your results from the preceding part to derive an equation for the local heat transfer coefficient, $h(x)$. You may assume that the latent heat of fusion, $\hat{h}_{sf}$, is much greater than any sensible heat $(c_p \, \Delta T)$ involved, that $\delta \ll b/2$, and that the heat transfer coefficient is defined as

$$h \equiv \frac{q_w}{(T_w - T_{mp})}.$$

4.6 Cooling water flows through a bundle of tubes in a heat exchanger. The tubes have an outside diameter $D = 2.0$ cm and are on a triangular pitch $P = 3.0$ cm (Fig. 4.18). The tube bundle is contained in a cylindrical shell, with the outermost tubes' surfaces lying 2.0 cm from the vessel wall. The overall length of the bundle is 7 m and the vessel has an interior diameter of 1 m.

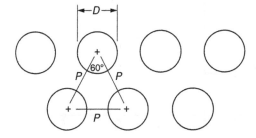

Figure 4.18. Layout of tubes in tube bundle.

a. If the water outside the tubes flows parallel to the axes of the tubes with a bulk velocity of $u_b = 0.90$ m/s, estimate the heat transfer coefficient between the water and the tubes for tubes near the interior of the bundle.

b. Estimate a heat transfer coefficient for tubes near the vessel wall, assuming that the bulk velocity is the same at those positions as in the center of the bundle.

c. Now suppose that the baffles are placed within the bundle, forcing the flow to travel from top to bottom, shown in Figure 4.19. If the baffles are located 1 m apart and stand 0.66 m from the bottom (or top) of the vessel, estimate the heat transfer coefficient. Assume that the total mass flow of water is unchanged.

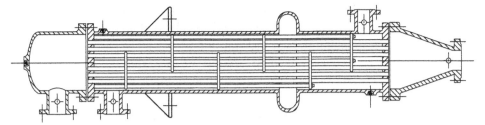
Figure 4.19. Tube bundle with baffles [4.25].

4.7 For laminar flow over a flat plate with constant wall temperature, do a physically based qualitative assessment of the influence of the following situations on the wall to fluid heat flux, $q_w(x)$, and the heat transfer coefficient, $h(x)$:

   a. The plate has a large number of very small holes through it and there is suction applied so that a modest amount of fluid is sucked through the plate at a low velocity normal to the plate.

   b. With the same geometry, a pressure is applied to the underside of the plate so that a modest amount of fluid at ambient temperature passes through the plate at low velocity and joins the flow above the plate.

   c. The plate is separated into a series of N smaller plates each with a length of L/N. The gap between plates is $1/4\,L/N$.

4.8 Fluid passes over a plate with constant wall temperature in laminar flow. A pressure gradient is applied in the flow direction (nonzero dp/dx). At the leading edge of the plate, the fluid has a uniform velocity U_0 equal to the velocity for the case of zero pressure gradient. Compare $q_w(x)$ and $h(x)$ between the zero pressure gradient case and the following:

   a. The pressure increases along the plate length, a positive dp/dx (upstream of any boundary layer separation)

   b. The pressure decreases along the plate length, a negative dp/dx.

4.9 Verify the results given in Eq. (4.33) for laminar and turbulent flows. For laminar flow, assume a circular tube of radius R with a Poiseuille velocity profile: $u(r) = 2u_b[1 - (r/R)^2]$. For the case of a turbulent flow, velocity profile in a circular pipe may be approximated by the formula $u/u_b = (n+1)(2n+1)(y/R)^{1/n}/(2n^2)$, for y the distance from the wall, R the pipe radius, and n a number ranging from roughly 6 to 10 (see [4.26, Section XX.a]). A correlation for n as a function of Reynolds number is [4.27]: $n = \ln \text{Re}_D/(1.085 + 6.535/\ln \text{Re}_D)$.

4.10 Your boss has suggested that half-moon shaped tubing will not only have improved heat transfer coefficients, but will also be good for marketing: "stellar performance," "space-age technology," and "heat transfer rates out of this world." Estimate the ratio of Nusselt number for a half moon shaped duct to that for a circular duct for both laminar and turbulent flow. Compare your laminar flow estimate to an exact value from literature. What will you tell your boss about his suggestion?

4.11 In a circulating fluidized bed, gas moves rapidly up a vertical column. The drag force of the gas balances the weight of a dense cloud of modest sized solid particles (particle diameter between 100 μm and 500 μm). The particles form loose clusters that are actively mixed. Measurements show that the convective heat transfer from the surface of the column is increased by an order of magnitude compared to single-phase gas flow at the same gas velocity. There are two possible explanations for the enhancement of the transfer:

a. The cluster of particles move along the wall frequently renewing the thermal boundary layer of the gases adjacent to the wall, similar to the action that rib roughness on a wall has in repeatedly interrupting the boundary layer; or

b. The clusters increase the heat capacity of turbulent eddies that move laterally between the wall and the center of the column.

To determine which of these mechanisms is most important, a series of experiments based on the analogy of heat and mass transfer were carried out. In single-phase gas flow, without the particles, close agreement was found between the Nusselt number for heat transfer and the Sherwood number for mass transfer. When the bed was operated at the same velocity with sand particles fluidized by the gas, the increase in Nusselt number was an order of magnitude larger than the increase in Sherwood number. Which of the two explanations given above do you think is responsible for this disparity?

4.12 High-voltage underground transmission lines have three large cables inside a 0.5 m diameter pipe as shown in Figure 4.20. Helical wires are wound around the outer surface of each cable to aid in drawing the cables through the pipe and to provide separation between adjacent cables. The space between the cables and the pipe is filled with a dielectric fluid that provides cooling to absorb the I^2R losses. The cables are usually 30 K above the bulk temperature of the fluid. In the conventional design, heat is transferred from the pipe to the surrounding soil. In the new design, the dielectric fluid will be chilled and pumped through the pipe to provide enhanced cooling and therefore higher allowable current transmission levels. To experimentally determine the approximate contribution of natural convection and forced convection in this new design, you are asked to outline one or more experiments using the dielectric fluid. Can you think of full-scale or properly scaled experiments that can be done without requiring a long, full-scale pumped system and that can be carried out under reasonable experimental conditions?

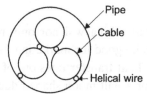

Figure 4.20. High-voltage transmission cable cross section.

4.13 Figure 4.21 shows some experimental results and theoretical predictions for heat transfer from particles in a packed bed to fluid flowing through the bed.

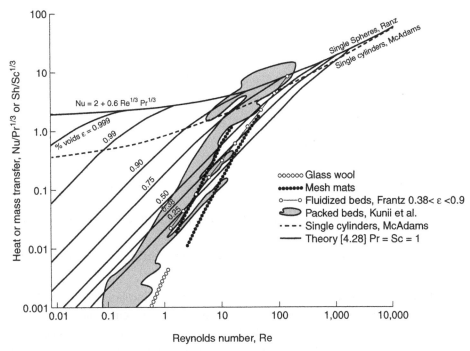

Figure 4.21. Heat transfer in packed beds. Theory shown by solid lines. Nu and Re are based on particle diameter [4.28].

The Nusselt number, Nu, is the particle to fluid heat transfer rate per particle surface area based on the local particle temperature and the local bulk fluid temperature. Consider the predictions with low fluid volume fraction, or void fraction, at low Reynolds numbers. Do these seem reasonable when compared to the lower limit for laminar heat transfer in internal flows?

In the experiments, the particles are packed into a large tube. They are heated to a constant temperature. Cold air then flows through the bed and the heat transfer rate is found from the change of the bulk air temperature between air inlet and outlet. Can you think of any experimental conditions that would contribute to these very low Nusselt numbers?

4.14 In an experiment on natural ventilation in a closed room, one end wall was heated and the other end wall was cooled to a uniform temperature. All other walls and the ceiling and floor were insulated. The average temperature in the center of the room was 20°C while the cold wall was at 15°C. It was observed that a heated jet formed along the hot wall, traveled along the ceiling, and then moved down the cold wall. The jet was wide and had a temperature of 25°C and an average velocity downward of 0.7 m/s at the top of the cold wall. Part way down the wall the jet divided. The region close to the cold wall continued down while the region of the jet farther from the wall separated and reversed direction as shown in Figure 4.22.

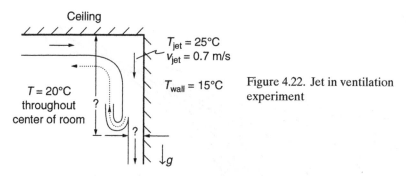

Figure 4.22. Jet in ventilation experiment

a. Explain qualitatively the physics of the observed flow. Why did the reverse flow occur?

b. Estimate the vertical location down the cold wall where the flow reversal occurred.

c. Estimate the thickness of the region that remained at the wall when the division occurred.

4.15 In deflagrations, or "slow combustion," a gas mixture burns as a low speed flame moves through it. The burning process is a chemical reaction that converts the reactants to products, releasing heat. The speed of the flame front is governed by a balance between the kinetics of the reaction and the diffusion of heat from the flame into the cooler gas ahead of it. The reaction rate is strongly dependent on the temperature.

As a simple model of this process, neglecting mass transfer, consider a planar flame front that advances into stationary, premixed reactants at T_∞. The flame front has a constant speed V_f and leaves reactants at a temperature T_0 behind it (Fig. 4.23). Assume that all thermophysical properties are constant and known.

a. If the reaction releases heat Δh^o (J/kg) per unit mass of reactants, give an expression for the temperature of the products.

b. Assume that the heat of reaction is released at the hottest point in the flame, where the gas reaches T_0. Heat flows from that point into the unburnt gas

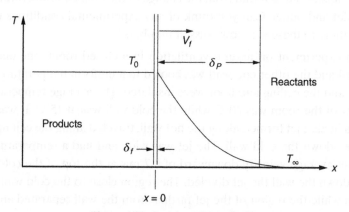

Figure 4.23. Laminar flame front.

ahead of the flame, creating a preheated zone of width δ_p. In the preheat zone, the cold reactants are warmed from T_∞ to T_0. Estimate the temperature distribution in the preheat zone and the thickness δ_p in terms of the flame speed V_f and physical properties.

c. The reaction rate increases exponentially with temperature, so it is not unreasonable to assume that the reactants are consumed mainly when the mixture is near T_0. The region where this occurs is then a thin flame of thickness $\delta_f \ll \delta_p$. If the time for the reaction to go from start to finish at this temperature is τ_0, find δ_f in terms of τ_0 and V_f.

d. Since the reaction actually begins at temperatures below T_0, a mean reaction time for temperatures between T_∞ and T_0 can alternatively be considered. Call this time $\bar{\tau}$. Owing to the effect of temperature on the reaction rate, $\bar{\tau} \gg \tau_0$. If the overall flame thickness is $\delta = \delta_f + \delta_p$, find V_f in terms of $\bar{\tau}$ and physical properties.

e. The ratio of $\bar{\tau}$ to τ_0 is the Zel'dovich number, $\mathrm{Ze} = \bar{\tau}/\tau_0$. Write the V_f, δ_f and δ_p in terms of Ze, τ_0, and physical properties.

5

Heat Exchangers

5.1 Introduction

Heat exchangers transfer heat from one fluid stream to another. They are used in power plants, vehicles, air conditioners, computer data centers, and a wealth of other applications. The fluids may both be liquids, both be gases, or be a combination of the two. The exchangers increase the overall energy conversion efficiency of power and propulsion applications. In some instances, heat exchangers involve simultaneous heat and mass transfer, as in evaporative cooling towers and so-called swamp coolers used to chill air in hot, dry climates. Heat exchanger problems usually involve either the performance prediction of an existing device (sometimes called "rating") or the design of a new heat exchanger for a given set of performance goals. Calculations of heat exchanger performance bring in principles of conduction for extended or finned surfaces and of convection either inside or outside tubing. For high-temperature applications, thermal radiation can also be important.

In this chapter, we focus on general principles of heat exchanger operation, the influence of geometry, and limiting performance levels. These ideas will be applied to make basic estimates of actual or limiting performance.

5.2 Heat Exchanger Geometry

Heat exchangers usually consist of an array of tubes, plates, or other solid surfaces that physically separate the hot stream from the cold one while allowing heat transfer between the two. A heat exchanger is shown schematically in Figure 5.1. Hot fluid enters at T_{Hin} and leaves at T_{Hout}. The cold stream enters at T_{Cin} and exits at T_{Cout}. The internal geometry of the heat exchanger is usually arranged to assure uniform flow distribution of both streams over their respective cross-sectional areas.

The flow arrangement may have the two fluid streams flowing parallel in the same direction or in opposite directions (counterflow). In other designs the two fluids streams may flow at right angles (crossflow), or they may have a combination of these, such as found in a shell and tube heat exchanger. Figure 5.2 illustrates several examples: a compact exchanger in crossflow, a shell and tube exchanger in counterflow, and a plate and frame exchanger.

156

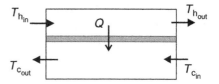

Figure 5.1. Schematic diagram of a heat exchanger.

5.3 Energy Balance, Limiting Cases

Usually, conditions in a heat exchanger are in steady state, although devices such as a rotary heat exchanger have periodic conditions in which the solid heat exchanger material alternatively contacts the cold and then the hot fluid streams.

A simple estimate of the upper limit on heat transfer can be obtained from an energy balance for one fluid in steady state. The heat transfer to the hot stream is

$$Q_H = -(\dot{m}_H c_{pH})(T_{H\text{in}} - T_{H\text{out}}) = -C_H(T_{H\text{in}} - T_{H\text{out}}) \tag{5.1}$$

And the heat transfer to the cold stream is

$$Q_C = (\dot{m}_C c_{pC})(T_{C\text{out}} - T_{C\text{in}}) = C_C(T_{C\text{out}} - T_{C\text{in}}) \tag{5.2}$$

where C_H and C_C are the products of mass flow rate and specific heat for the hot and cold stream, respectively. They are referred to by several different names in the literature, for example, as the heat capacity rates. We refer to them herein as heat capacity. This formulation assumes neither fluid undergoes a phase change and that the specific heat of the two streams is constant.

Neglecting kinetic energy, the upper limit for heat transfer occurs when one of the two fluids, say the hot fluid, is cooled to the entering temperature of the other (cold) fluid. By the second law of thermodynamics, the hot fluid temperature cannot be any cooler than this. This gives an upper limit on the heat transfer to the hot stream:

$$Q_{\text{upper limit}_H} = -(\dot{m}_H c_{pH})(T_{H\text{in}} - T_{C\text{in}}) = -C_H(T_{H\text{in}} - T_{C\text{in}}) \tag{5.3}$$

An alternative limit is obtained by applying this limiting temperature difference to the cold stream:

$$Q_{\text{upper limit}_C} = (\dot{m}_C c_{pC})(T_{H\text{in}} - T_{C\text{in}}) = C_C(T_{H\text{in}} - T_{C\text{in}}) \tag{5.4}$$

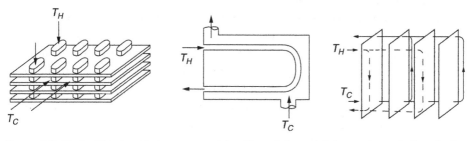

Figure 5.2. Typical heat exchanger geometries (from left to right): compact, shell and tube, and plate and frame heat exchanger.

Heat transfer from the heat exchanger to the environment surrounding it is often negligible in comparison to the heat exchange between the two streams. Thus, the energy gain of the cold stream must be equal in magnitude to the energy loss of the hot stream:

$$-C_H \Delta T_H = C_C \Delta T_C \tag{5.5}$$

Since C_H usually does not equal C_C, it is clear that one of the limits given by Eqs. (5.3) and (5.4) will be smaller than the other. If $C_H > C_C$, Eq. (5.5) shows that the hot stream must have a smaller temperature change than the cold stream. Further, if the cold stream has undergone the maximum possible temperature change of $T_{Hin} - T_{Cin}$, then the temperature change of the hot stream cannot be as large. Thus, for $C_H > C_C$, only Eq. (5.4) based on the cold stream heat capacity rate is a valid upper bound. When $C_c > C_H$, the conclusion is opposite: Eq. (5.3) based on the hot stream heat capacity rate is the valid upper limit. In general, then, the upper limit should be expressed as

$$Q_{\text{upper limit}_{C_{\text{MIN}}}} = C_{\text{MIN}}(T_{Hin} - T_{Cin}) \qquad \text{(High NTU limit)} \tag{5.6}$$

where C_{MIN} is the minimum of C_H or C_C. (NTU is explained shortly.)

An alternate expression for the upper limit of heat transfer can be given in terms of the overall heat transfer coefficient, U, between the hot and cold stream and the surface area for heat transfer, A. The maximum possible heat transfer would occur if the temperature difference for heat transfer remained at its maximum level over the entire surface area of the exchanger, that is, if the hot stream remained at its inlet temperature and the cold stream remained at its inlet temperature:

$$Q_{\text{upper limit}UA} = UA(T_{Hin} - T_{Cin}) \qquad \text{(Low NTU limit)} \tag{5.7}$$

There remains the issue of which of the upper limit expressions, Eq. (5.6) or (5.7), is a realistic estimate. Dividing Eq. (5.7) by Eq. (5.6), the ratio of the two upper limit expressions is given by the dimensionless parameter, UA/C_{MIN}, referred to as the number of transfer units or NTU:

$$\frac{Q_{\text{upper limit}_{UA}}}{Q_{\text{upper limit}_{C_{\text{MIN}}}}} = \frac{UA}{C_{\text{MIN}}} = NTU \tag{5.8}$$

For high heat capacity C_{MIN} and low UA product, that is, a low NTU value, the temperatures of the hot and cold streams would not change much and Eq. (5.7) would be a reasonable upper limit to use. This situation maybe visualized, for example, as a very short heat exchanger in which heat transfer is minimal relative to the enthalpy rates of the fluid streams. In the opposite extreme of very high values of UA relative to C_{MIN}, a high NTU value, the temperatures of the hot and cold streams would change appreciably. In this case, Eq. (5.7) would be a poor estimate, and Eq. (5.6) would be preferable. In fact, Eq. (5.6) represents the limit attained when the UA product becomes very large and the flow geometry through the exchanger is also optimal so that the streams reach the thermodynamic upper limit of heat exchange.

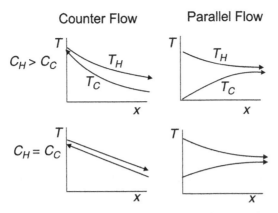

Figure 5.3. Limiting cases for counter and parallel flow when UA/C_{MIN} becomes large.

Figure 5.3 illustrates the limiting temperature patterns for counter flow and parallel flow heat exchangers when UA/C_{MIN} becomes very large. Note that the geometry of the flow and the ratio of C_{MIN}/C_{MAX} come into play in determining the actual upper limit of performance. Consider the case of a hot stream and cold stream with equal values of heat capacity and both streams flowing in the same direction (parallel flow). As the hot stream experiences a temperature drop, the cold stream experiences a corresponding temperature rise. In the limit, both streams will approach the same temperature, which will be the average of the inlet hot and cold temperatures. Both streams approach only one half of limiting temperature change, $T_{Hin} - T_{Cin}$.

When the two streams are flowing in opposite directions, termed counterflow, the cold stream outlet is adjacent to the inlet of the hot stream. In counterflow, when the cold stream has the minimum heat capacity the cold stream temperature will approach the inlet hot stream temperature as the area gets very large. The counterflow yields a higher net heat transfer than the parallel flow or any other flow geometry for the same ratios of UA/C_{MIN} and C_{MAX}/C_{MIN}. Thus, the limit given by Eq. (5.6) applies to counterflow, and it is also the upper limit for any heat exchanger geometry. In the special case of counterflow with $C_H = C_C$, called "balanced flow," both streams can approach a temperature change of $T_{Hin} - T_{Cin}$, when the NTU is very large.

5.4 Heat Exchanger Performance Relationships

Calculation of the heat exchanger size required to transfer a certain magnitude of heat between two streams requires a combination of heat transfer and thermodynamics. The overall heat transfer coefficient for exchange between two fluid streams separated by a solid wall of thickness W is

$$U = \left[\frac{1}{h_H} + \frac{W}{k} + \frac{1}{h_C} \right]^{-1} \tag{5.9}$$

This expression does not account for the presence of extended surfaces, such as fins, on either hot or cold side and assumes that the wall has the same area on each side (i.e., it is thin relative to any radius of curvature).

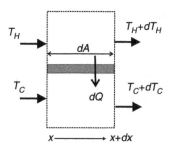

Figure 5.4. Differential area for heat transfer from hot to cold stream.

The heat transfer from the hot to the cold stream for a differential area dA can be written as

$$dQ = U\, dA(T_H - T_C) \tag{5.10}$$

In this expression T_H and T_C represent the local bulk temperature of the hot and cold stream, respectively, at the coordinate x where dA is located.

We can also write an energy balance for the hot and cold streams by taking a control volume over each of the streams from distance x to $x + dx$ (see Fig. 5.4). For the cold air stream with steady-state conditions the heat added is

$$dQ = C_C dT_C \tag{5.11}$$

and for the hot air stream, the heat added is

$$dQ = -C_H dT_H \tag{5.12}$$

Proceeding in the x direction, the cold stream temperature rises while the hot stream temperature falls. (Note that the same energy balance applies when the fluids are in counterflow, but the temperatures of *both* streams will either rise in the x direction or both will fall in the x direction.) For good energy efficiency, it is useful to transfer as much energy from the hot to the cold stream as possible. Then T_C should approach the hot stream temperature, T_H. The question is: How large must the surface area of the heat exchanger be to achieve a given cold stream temperature rise? The difficulty in calculating the area is that T_H and T_C both vary with x. In the general case, Eqs. (5.10), (5.11), and (5.12) must be solved simultaneously.

Rather than starting with the most general case, consider instead a situation that will yield a better estimate for the upper limit of the actual heat transfer rate. If we wish to find the maximum heat transfer for a fixed cold stream flow rate and a fixed exchanger area, we would like to maximize the cold stream temperature increase. To accomplish this, we want the hot stream temperature to stay close to its maximum inlet temperature throughout the exchanger. That will occur when the hot stream has a much higher heat capacity than the cold air stream. Comparing Eqs. (5.11) and (5.12) in this limit, we see that the hot stream temperature remains nearly constant over the entire length while the cold stream temperature continuously increases in x. (One example of this occurs when the hot stream is a saturated vapor condensing at constant pressure over the length of the exchanger.) Combining Eqs. (5.10) and

(5.11), since dQ is the same in both expressions, gives

$$C_C dT_C = U dA (T_H - T_C) \tag{5.13}$$

Since T_H has been taken to be constant for this case, we can rearrange this as

$$\frac{d(T_H - T_C)}{(T_H - T_C)} = -\frac{U dA}{C_C} = -\frac{U dA}{C_{MIN}} \tag{5.14}$$

Integrating Eq. (5.14) yields

$$\frac{(T_C - T_H)_{x=L}}{(T_C - T_H)_{x=0}} = \frac{(T_{Cout} - T_H)}{(T_{Cin} - T_H)} = e^{-\frac{UA}{C_{MIN}}} \tag{5.15}$$

where T_{Cin} is the entering cold stream temperature (at $x = 0$) and T_{Cout} is the leaving temperature (at $x = L$). As the area is increased, the cold stream leaving temperature approaches the hot stream temperature. Note the diminishing returns with increasing area A as the magnitude of the exponent on the right-hand side of Eq. (5.15) becomes larger.

The more usual representation of heat exchanger performance is in terms of the effectiveness, which is defined as the ratio of actual heat transfer to the limiting value in counter flow at large UA while the flow rate remains fixed. That limiting value is set by Eq. (5.6). The effectiveness, ε, becomes,

$$\varepsilon = \frac{Q_{actual}}{Q_{limit\ as\ A \to \infty}} = \frac{C_C(T_{Cout} - T_{Cin})}{C_{MIN}(T_{Hin} - T_{Cin})} \tag{5.16}$$

For this example, C_C is the minimum heat capacity and T_H is assumed to remain constant, so we can rewrite Eq. (5.16) by adding and subtracting T_H from the two terms in the numerator. Then, using the solution from Eq. (5.15), we have

$$\varepsilon = \frac{(T_{Cout} - T_H) - (T_{Cin} - T_H)}{(T_{Hin} - T_{Cin})} = \frac{(T_{Cin} - T_H)\left[e^{-\frac{UA}{C_{MIN}}} - 1 \right]}{(T_{Hin} - T_{Cin})} \tag{5.17}$$

Writing the dimensionless ratio UA/C_{MIN} as NTU, the final form of Eq. (5.17) is

$$\varepsilon = 1 - e^{-NTU} \tag{5.18}$$

This relationship is shown in Figure 5.5 as the case of $C_{MIN}/C_{MAX} = 0$ and is referred to as the single stream limit. Note that, for this limiting case, the geometry of the exchanger does not influence the effectiveness.

Extending the above analysis to nonzero values of C_{MIN}/C_{MAX}, now T_H decreases as the hot fluid flows through the exchanger. At any location, the hot to cold temperature difference is decreased reducing the overall effectiveness. In the general case when the heat capacity of both streams is the same magnitude, the effectiveness, ε, is a function of the NTU, but it is also a function of the specific flow geometry and the ratio of heat capacity between the two streams. Figure 5.5 shows the effectiveness NTU relationship for counterflow.

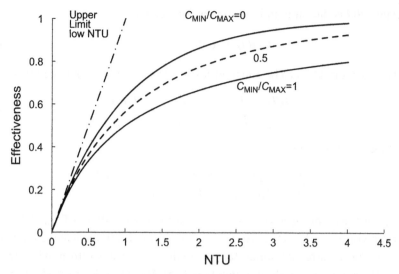

Figure 5.5. Effectiveness-NTU relationship of counterflow exchanger.

For small values of NTU, it was pointed out previously that Eq. (5.7) would be the correct upper limit. Substituting into the definition of the effectiveness yields

$$\varepsilon_{\text{upper limit}} = \frac{Q_{\text{upper limit}}}{Q_{\text{limit as } A \to \infty}} = \frac{UA(T_{H\text{in}} - T_{C\text{in}})}{C_{\text{MIN}}(T_{H\text{in}} - T_{C\text{in}})} = NTU \quad \text{when NTU} \ll 1$$

(5.19)

The limit for large NTU is given by Eq. (5.6). This limit is not very useful because it indicates an upper limit effectiveness of unity. A more useful upper limit for all exchanger geometries is set when $C_{\text{MIN}}/C_{\text{MAX}}$ approaches 0, Eq. (5.18).

For an exchanger with a known geometry, $C_{\text{MIN}}/C_{\text{MAX}}$ and UA values, the effectiveness can be found from analytical relationships or from graphs of ε versus NTU as a function of $C_{\text{MIN}}/C_{\text{MAX}}$. For counterflow configurations, the general relationship is [5.1]:

$$\varepsilon = \frac{1 - e^{-NTU(1-C_{\text{MIN}}/C_{\text{MAX}})}}{1 - (C_{\text{MIN}}/C_{\text{MAX}})e^{-NTU(1-C_{\text{MIN}}/C_{\text{MAX}})}}$$

(5.20)

The overall heat transfer rate can then be found, by using Eq. (5.16), to be:

$$Q = \varepsilon C_{\text{MIN}}(T_{H\text{in}} - T_{C\text{in}})$$

(5.21)

Figure 5.6 shows the effectiveness for a crossflow exchanger where both fluids are unmixed, that is, when there is no lateral mixing of the hot or cold fluid across their respective cross-sectional areas. The compact exchanger shown in Figure 5.2 is an example of a crossflow exchanger. For all values of $C_{\text{MIN}}/C_{\text{MAX}}$ greater than 0, the counterflow arrangement yields a higher effectiveness than the crossflow.

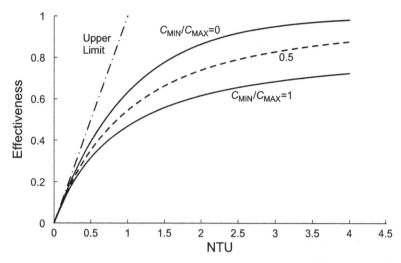

Figure 5.6. Effectiveness-NTU relationship for crossflow exchanger with fluids unmixed.

5.5 Heat Exchanger Design

Heat exchanger design or selection involves a number of different considerations. The designer must trade off thermal performance with fluid pressure drop, exchanger cost, size, and weight, among other factors [5.2]. For example, to minimize pressure drop, the cross-sectional frontal areas for flow can be enlarged to reduce velocity. This, in turn, may reduce convective heat transfer coefficients and perhaps create awkward dimensions. Numerous heat exchanger surface geometries have been developed, some with the goal of cost reduction, others emphasizing performance improvements. The latter aim to enhance heat transfer while minimizing increases of the friction factor [5.3]. For example, adding transverse ribs to a smooth surface will increase the convective heat transfer coefficient; aligning the ribs at an acute angle to the flow direction increases the ratio of heat transfer to frictional loss [5.4].

Detailed design of the exchanger must await resolution of the overall trade-off issues. These trade-offs are another instance in which an initial, qualitative consideration of the limiting conditions can help guide the design. Example 5.1 illustrates one such issue.

Example 5.1 *Liquid coupled heat exchangers* In large commercial buildings, considerable energy can be saved in the winter by using the warm exhaust air from the building to preheat cold incoming ventilation air. To avoid reentrainment of the exhaust into the air inlet, the inlet and exhaust locations must be kept well apart. To facilitate this separation, it may be easier to transfer the heat from the exhaust to the inlet stream by use of an intermediate liquid loop as shown in Figure 5.7. Heat is transferred from the exhaust air to the liquid in one exchanger. The warm liquid is circulated to a second exchanger, where it raises the temperature of the incoming air stream. If the flow rate and heat capacity of the exhaust and incoming air streams are approximately the same, what should the flow rate

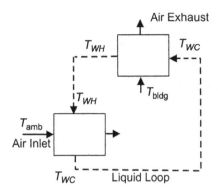

Figure 5.7. Heat recovery loop for a building. Winter operation.

of the liquid loop be to maximize overall effectiveness of the combined system? Assume that the two exchangers are both crossflow and that Figure 5.6 can be used to judge their performance.

An initial thought is to aim for high effectiveness in each heat exchanger by increasing the mass flow rate of the liquid loop. This would result in a liquid heat capacity much greater than that of the air streams, causing C_{MIN}/C_{MAX} to be small and raising the effectiveness of each exchanger for a given UA/C_{MIN}. To evaluate this idea, we will look at limiting solutions combined with an energy balance for the system. For the liquid loop,

$$Q = C_{\text{liquid}}(\Delta T_{\text{liquid}}) \tag{5.22}$$

When C_{liquid} is much greater than C_{air}, ΔT_{liquid} is much smaller than ΔT_{air}. Then the overall temperature change for the liquid stream and the two air streams will follow the pattern shown on Figure 5.8a, where the flow rate of inlet and exhaust air streams are the same. Clearly, even when the NTU of the two exchangers is very large, the temperature rise of the inlet air stream is, at most, one half of the overall temperature difference between the ambient air and the building interior air temperature. Thus, in this extreme case of high liquid heat capacity, the overall effectiveness of the entire system is at best 0.5. This example is typical of situations in which the independent optimization of subsystems does not optimize the overall system.

Taking the other extreme of very low liquid flow rate, C_{MIN}/C_{MAX} of each exchanger again will approach 0, giving a high effectiveness. However, from a simple energy balance on the liquid stream, Eq. (5.22), when C_{liquid} is very small, the overall heat transfer is also small. In this case, the effectiveness and NTU are based on the liquid heat capacity while we are interested in the effectiveness of heat transfer to the air stream. The optimal value of the liquid heat capacity rate is not at *either* extreme value. It can be shown [5.1] that the optimum occurs when the heat capacity of the liquid stream is equal to the air stream heat capacity rate (i.e., when both heat exchangers are balanced). Results for this case are illustrated in Figure 5.8b.

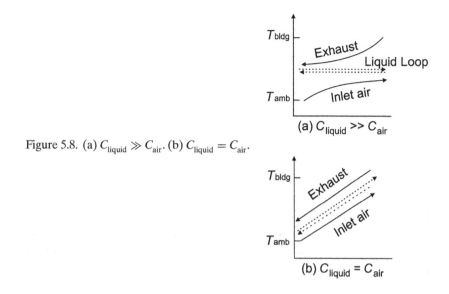

Figure 5.8. (a) $C_{\text{liquid}} \gg C_{\text{air}}$. (b) $C_{\text{liquid}} = C_{\text{air}}$.

Example 5.2 *Nonuniform flow* Designers strive to design heat exchangers to have uniform flow over the inlet cross section. For example, consider a cooler that has hot liquid passing through a series of parallel tubes. The cooling fluid flows outside of the tubes in a counterflow arrangement as shown in Figure 5.9. The heat capacity of the cooling fluid is twice that of the hot fluid. Consider the case of flow maldistribution, so that the hot fluid mass flow is not split equally between the upper and lower tube bundle. Instead, the flow rate through the top bundle is increased by a reduction of flow rate through the bottom bundle, while the combined flow rate remains the same. As a result, the overall performance, measured in terms of the total heat removed from the combined tube bundles, suffers.

We can consider the upper and lower tube bundles, and associated hot and cool streams, as two separate heat exchangers. They both have the same heat transfer area but with different values of C_{MIN}. Thus, the top bundle with the increased flow rate has a higher C_{MIN}, a $C_{\text{MIN}}/C_{\text{MAX}}$ ratio closer to unity, and a lower NTU, while the bottom bundle has an increased NTU. We now have a

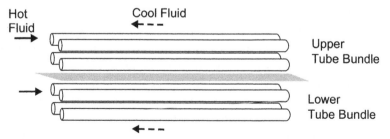

Figure 5.9. Cooler with separate tube bundles. The hot fluid is inside the tubes and the cool fluid is in counterflow outside the tubes.

majority of the fluid passing through an exchanger, the top bundle, which has a lower NTU and a lower effectiveness. Remember that the total heat transfer rate is proportional to the product of effectiveness and C_{MIN} [see Eq. (5.21)]. Summing the heat transfer to the upper and lower tube bundles, the overall effect is a reduction in total heat transfer. Problem 5.6 explores this issue further.

An interesting example of the impact of flow imbalance occurs in an oil cooler. Hot oil flows inside tubes with cooling water flowing over the exterior tube surface. The viscosity of oil varies strongly with its temperature. If there is a small flow imbalance so that the oil in the top bundle has a higher flow rate, its average temperature through the length of the tube will increase. The increased temperature in the upper bundle results in a lower average viscosity while the oil in the lower bundle with a lower flow rate will become colder. Thus, the flow resistance in the upper bundle decreases while the resistance of the lower bundle increases. This will tend to promote a still larger imbalance with higher flow in the upper bundle, resulting in a growing instability [5.5].

Example 5.3 *Glass fiber spinning* In Chapter 1, the process of spinning glass fibers was described and some simple order of magnitude estimates were made for the required air flow rate to ensure uniform cooling of all of the fibers. Now, a more careful estimate of uniformity can be made. The cooling process is basically a crossflow heat exchanger with the cooling air flowing horizontally over the vertically flowing glass fibers. The heat capacity of the air flowing over the fibers is much higher than the heat capacity of the flowing glass. This ensures that the average air temperature increase is minimal over the process.

The temperature change of the glass fibers closest to the cooling air inlet can be determined by considering a crossflow exchanger involving the air flow at its delivery temperature along with the first group of fibers closest to the entering air. For these fibers, UA/C_{MIN} is based on the flow rate and surface area of only the first group, while C_{MAX} is the based on the flow rate of the entire air flow. Because the air is coolest for this bunch of fibers, these fibers should achieve the lowest final temperature.

We can compare this to the average temperature decrease for all of the fibers. Consider the entire group of fibers and the crossflow air as another, larger crossflow exchanger. Again using the crossflow effectiveness–NTU relationship, we can find the total heat transfer and temperature decrease for entire group of fibers. Note that the value of UA/C_{MIN} is now based on the flow rate and surface area of all of the fibers; it should have the same value as the UA/C_{MIN} for the first fiber bunch. However, the ratio of heat capacities, C_{MIN}/C_{MAX}, will be substantially increased when considering all of the fibers. The ratio of the temperature decrease for the first group of fibers to the average for all of the fibers gives us a measure of the uniformity of the process. Because the fibers must be cooled to a temperature near ambient a large value of the NTU is required. We will take it to be 3.0. The heat capacity ratio of fibers to air will be modest; for this example we will use 0.25 as the overall ratio C_{MIN}/C_{MAX}. For the first group of fibers C_{MIN}/C_{MAX} is close to 0. Using the effectiveness NTU relationship for crossflow given by Kays and London [5.1], the ratio of temperature decrease of the first group to the average for all of the fibers is 1.07. In using the crossflow solutions, we need to assume that the air flow is unmixed in the vertical plane. We could also examine the results at the other limit, assuming that the air flow is completely mixed in the plane normal to its direction of flow. In the latter case, the temperature change is reduced by 12%. The latter would represent the a lower limit to the glass temperature change.

REFERENCES

[5.1] W. M. Kays and A. L. London. *Compact Heat Exchangers*. New York: McGraw-Hill, 1964.
[5.2] A. P. Fraas. *Heat Exchanger Design*, 2nd ed. New York: John Wiley & Sons, 1989.
[5.3] A. E. Bergles. Techniques to enhance heat transfer. In W. M. Rohsenow, J. P. Harnett, and Y. I. Cho (eds.), *Handbook of Heat Transfer*. New York: McGraw-Hill, 1998.
[5.4] J. C. Han, L. R. Glicksman, and W. M. Rohsenow. An investigation of heat transfer and friction for rib-roughened surfaces. *International Journal of Heat and Mass Transfer*, **21**(8), 1143–1156, 1978.
[5.5] W. M. Rohsenow. Private communication to L. R. Glicksman, 1969.

PROBLEMS

5.1 A periodic or rotary heat exchanger uses a rotating wheel made up of a fine mesh or honeycomb made of metal or ceramic material. Hot gas passes through the upper part of the disk and cold gas passes through the lower part of the disk (see Fig. 5.10). Heat is transferred from the hot gas to the solid disk material, raising its temperature. The hot disk rotates to the cold side, where heat is transferred from the hot material to the cold gas flowing through it. The process is then repeated as the cool disk material is brought in contact with the hot gas again.

a. Similar to Figure 5.3, sketch the average hot gas and cold gas temperatures versus distance, z, through the disk. Also sketch the average disk temperature versus z when it is in contact with the hot gas and when it is in contact with the cold gas.

b. Should the hot and cold streams be in parallel or counter flow arrangement? Show the approximate temperature profiles from part a for both arrangements.

c. Show the disk temperature for the limiting case when the disk rotational speed is high. Does this represent the upper limit for overall heat transfer effectiveness for given disk size and air flow rates?

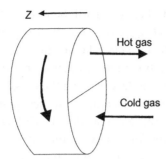

Figure 5.10. Rotary heat exchanger.

5.2 A heat pipe is proposed for transferring heat from a stream of hot gas to a stream of cold gas in an adjacent duct. The heat pipe is a closed tube containing a fluid in its saturated state. One end of the heat pipe is immersed in the hot gas and the other is in the cold gas. The fluid in the hot end of the heat pipe is evaporated at constant pressure due to heat transfer from the hot gas. The vapor flows to the other end of the tube, where the vapor is condensed at constant pressure by heat transfer to the cold gas. The condensate flows back to the hot side through a wick within the heat pipe. The temperature of the fluid throughout the heat pipe can be considered to be uniform.

  a. How does the limiting performance of the heat pipe compare to that of a rotary heat exchanger between the same two gas streams? Compare the effectiveness of these arrangements. What are other advantages of the heat pipe?

  b. A number of small diameter heat pipes will be used to exchange heat from the hot to the cold stream. Should they all be equally spaced at one cross section normal to the flow direction? Are there other arrangements that will improve performance?

5.3 A proposed new combustion process utilizes a vertical column to cool the combustion gases and to heat a flow of solid particles (see Fig. 5.11). Cool solid particles roughly 1 mm in diameter are introduced at the top of a tall vertical column and fall under the influence of gravity. Hot combustion gas is introduced at the bottom of the column and flows vertically upward at about 5 m/s. The proposed column is roughly 10 m high and between 1 m and 2 m in diameter. The solid particles enter at 100°C while the hot combustion gas enters the bottom of the column of at 1500°C. There are specially designed inlets and exits to separate the solid and gas at the two ends of the column. The mass flow rates of the solids and the gas are comparable.

  a. Develop an overall model for the gas-to-solid heat transfer from the top to bottom of the column. What are the governing mechanisms of heat transfer?

  b. How would you estimate the heat transfer coefficient from the solid surface to the gas?

  c. How do the flow characteristics of the solids influence the overall heat transfer? What would be some limiting cases for the behavior of the mass of solid particles as it moves down the column?

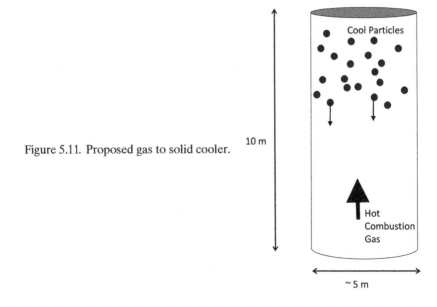

Figure 5.11. Proposed gas to solid cooler.

10 m

~ 5 m

5.4 In some homes, clothes dryers are installed near the center of the home. The dryer exhaust air is carried through a pipe in the basement or unheated space under the first floor (see Fig. 5.12). The exhaust air is at an elevated temperature, about 35°C and has a high relative humidity, near 70%, during most of the drying process. The air also contains lint, made up of small fibers of cotton and other materials. Lint tends to stick to the inside of the pipe wall if the wall is wet. Over time, the lint can build up in the pipe and block much of the air flow. The pipe has a diameter of 10 cm and the air velocity is 2 m/s.

   In the winter, the unheated space has a temperature of 5°C. When the dryer is off, there is no air flow through the pipe. The pipe through the unheated space is 8 m long.

   a. How would you estimate the temperature change of the pipe in the unheated space near the dryer exit, $x \approx 0$, starting from the time the dryer is turned on?

   b. Using an approximate model of the system, sketch the temperature versus the length of the pipe in the unheated space at several different times after

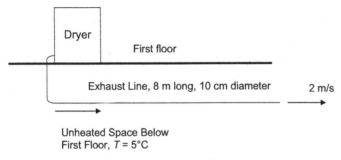

Figure 5.12. Dryer exhaust.

the dryer starts. Also estimate the water vapor concentration along the pipe length at different times.

c. Two choices for the pipe are lightweight aluminum or heavy galvanized steel. Which pipe material would do better to minimize lint build up?

5.5 An air-to-air heat exchanger is to be designed to preheat fresh air entering a house in winter. The house interior is 2000 ft.^{2} and is maintained at 70°F. The exterior air is at 40°F. One air change per hour (a volume flow rate of 2000 ft.^{3}/hr) is desired with the exhaust air preheating the incoming air. The incoming and exhaust air streams are separated by a series of thin parallel aluminum plates. The air velocity is 20 ft/s and the spacing between plates is 1 inch. What is the total area of plates necessary if the outside air is to be heated to 60°F? How can the exchanger be made more compact?

5.6 Consider the unbalanced flow in Example 5.2. Assume that the entire exchanger initially has an NTU of 1.0 and is counterflow. Assume the flow becomes unbalanced so that three fourths of the total hot fluid flows through the upper bundle, consisting of one half the total tubes, while the remaining one fourth of the hot flow is in the remaining tubes in the lower bundle. We would like to find the ratio of total heat transfer for this unbalanced exchanger to an equal sized exchanger with uniform distribution. Assume the change in flow rate does not substantially alter the overall heat transfer coefficient U. Do this for two cases. First, assume that the hot fluid has the minimum heat capacity rate and that C_{MIN}/C_{MAX} is very small. In the second case, assume that the heat capacity rate of the hot fluid matches that of the cold fluid when the flow is evenly distributed. Note that in both cases the cold fluid flow rate is equal for the upper and lower tube bundles.

5.7 In a gas fired boiler, saturated liquid water flows inside tubes and hot combustion gases flow over the outside of the tubes, so that the water leaves the tubes as saturated steam. Assume that the combustion process in the burner is completed before the gas contacts the tubes. The combustion gases enter the boiler at 850°C with a flow rate of 90 kg/s, and they can be assumed to have a constant specific heat of 1.5 kJ/kg K. The water is at 150°C and flows at 15 kg/s. The U value between the gas and water is 50 W/m^2 K. What is the required tube surface area for two limiting cases: (1) the gas flows along the length of the boiler in plug flow with no back mixing; and (2) the gas is well mixed at a uniform temperature throughout the interior of the boiler. Neglect radiation heat transfer.

5.8 Show that when a counterflow heat exchanger has equal hot and cold stream heat capacity rates, plots of the hot and cold steam temperature versus distance along the flow direction form parallel straight lines. Using the energy balances and heat transfer expressions, derive the effectiveness NTU relationship for this special case, which is known as a "balanced" counterflow heat exchanger.

6

Radiation Heat Transfer

6.1 Introduction

Thermal radiation is one of the two basic physical mechanisms of heat transfer. Energy is transferred from a warm body to a cooler one by electromagnetic waves or photons. Different disciplines take their own approach to thermal radiation. Physicists may focus on the implications of quantum mechanics. A material scientist interested in property values will use a microscopic point of view or Maxwell's equations. An engineer assessing heat transfer between finite objects will take a macroscopic point of view. The latter approach, employed in this chapter, utilizes basic phenomenological relationships to deal with complex engineering applications. The authors' experience is that the treatment of radiation heat transfer is sometimes shortchanged in introductory heat transfer classes. Therefore, in this chapter, we discuss the fundamental concepts as well as modeling and approximation.

Radiation is important in a wide variety of applications. In materials processing such as glassmaking, steel manufacture, and growth of electronic crystals, radiation has a first-order influence on resulting material properties. In the power industry, heat transfer from large combustors is primarily by radiation. In common thermal insulations, radiation is responsible for between one-fourth and one-half of the total heat transfer at room temperature. Radiation is similarly important in propulsion systems such as rocket nozzles and gas turbine blades. On a global scale, radiation is responsible for the heat balance of the earth, and changes in the radiative properties of the atmosphere are the cause of global warming.

The spectrum of electromagnetic radiation that is responsible for heat transfer spans from the visible wavelengths, 0.4 to 0.7 micrometers, through the near infrared, up to about 30 micrometers, and to the far infrared, from about 30 to 1000 micrometers. Because of the overlap of thermal radiation with the spectrum of visible light, some of the same geometric principles are used in both fields.

An engineer faces several difficulties when dealing with thermal radiation. The properties of solid surfaces governing radiation transfer are difficult to measure or predict. The properties are a function of the surface chemistry, the microscopic roughness, and the composition of the underlying material. In some instances the properties can change substantially with time, for example, when an initially shiny bare metal surface forms an oxide coating or becomes worn. The radiation absorbed and emitted by gases is dependent on the concentration of component species, such as CO_2. These properties are highly wavelength dependent.

A given body may receive thermal radiation from a number of surrounding bodies, so the spatial configuration to be analyzed can become complex, and interactions between multiple bodies may have to be considered simultaneously. In the most general case, the equations of transfer for radiation are written as nonlinear integrodifferential equations. Owing to the uncertainty in radiative property values for many cases, a precise solution of the equations of transfer is unwarranted, and the implied accuracy of such a solution is misleading. Much of the scientific literature dealing with accurate solutions rests on the assumption of highly simplified radiation property values.

Approximate methods of solution are appropriate in many of these instances when the property values are not well defined. Further, radiation at a point involves the integration over all directions of the flux from surrounding bodies; thus averaging over the boundary conditions and geometry may not cause substantial net errors.

6.2 Fundamental Concepts

The absorptivity, α, of a body is the fraction of the total incident radiant energy that the body absorbs. That is, the incident energy interacts with the matter of the body and the radiant energy is transformed to internal energy, to energy transfer by conduction, or to radiative energy that is emitted by the body in a different wavelength spectrum. A *black body* is defined as a body that absorbs all of the incident radiation and so has an absorptivity of unity.

All bodies at temperatures above absolute zero produce thermal radiation. The amount of radiant flux leaving a surface of a body is measured by the emissive power, e, in W/m^2. For a surface area dA, of a body, the energy emitted per unit time in all outward directions and all wavelengths is

$$dQ_{\text{emitted}} = e\,dA \tag{6.1}$$

Note this is the one-way flux *leaving* the body, not the net flux. For most solid bodies the energy leaving the surface is emitted by material very close to the surface and is characterized by the surface temperature. As a shorthand, the leaving energy is referred to as the energy emitted by a surface. For two bodies at the same temperature, it can be shown that a black body will have the highest emissive power, e_b. That emissive power is a function only of the temperature of the black body and the medium adjacent to it. The emissive power can be expressed as

$$e_b = n^2\sigma T^4 \tag{6.2}$$

where T is the absolute temperature of the body, n is the index of refraction of the medium adjacent to the body (for air, and for a vacuum, n is unity), and

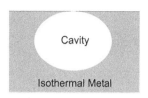

Figure 6.1. Black body cavity.

σ is the Stefan–Boltzmann constant, which has a value of approximately 5.67×10^{-8} W/m^{2}K^4 or 1.71×10^{-9} BTU/hr ft.^{2} °R^4.

A good approximation of a black body can be made by constructing a closed cavity in a metal. The metal should have a high thermal conductivity to ensure isothermal conditions (see Fig. 6.1). A small hole joining the cavity to the outside allows incident radiation to enter the cavity. If the hole is small compared to the cavity dimensions and if the cavity walls are all concave, then the amount of entering radiation that is reflected back out of the hole is very small. This makes the hole a near perfect absorber. To an observer on the outside, the area of the hole looks like the surface of a black body. This is true even though the absorptivity of the metal surfaces making up the cavity is less than unity.

If a true black body surface is outside the hole, facing the cavity, and is at the same temperature, then the net radiation transfer, the difference between outgoing and incoming flux, across the hole must be zero. That is, the flux leaving the cavity has the emissive power of a black body at the same temperature. Note that, no matter where the hole is located along the cavity surface, the radiant energy emitted from within will be the same. This idea can be extended to show that the incident radiation striking the surface of a small body placed within the cavity will be black body radiation.

6.3 Monochromatic Properties

Analogous to Eq. (6.1), we can describe the radiation energy leaving the surface of a black body within the wavelength range λ to $\lambda + d\lambda$ as

$$dQ = e_{b\lambda} dA d\lambda \tag{6.3}$$

and

$$e_b = \int_0^\infty e_{b\lambda} d\lambda \tag{6.4}$$

where the black body monochromatic emissive power, $e_{b\lambda}$, is a function of the temperature, the wavelength, and the index of refraction of the adjacent body. From quantum theory, Planck derived an expression that matched experimental measurements of the black body spectrum. When n is unity, his result is

$$\frac{e_{b\lambda}}{T^5} = \frac{2\pi C_1}{(\lambda T)^5 [e^{C_2/\lambda T} - 1]} \tag{6.5}$$

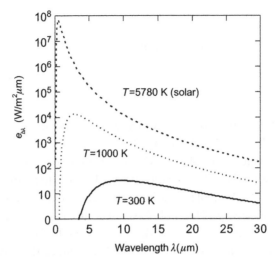

Figure 6.2. Monochromatic emissive power of a black body, $e_{b\lambda}$, as a function of wavelength and temperature.

Note, when written in this form the ratio of $e_{b\lambda}/T^5$ is a function of the product λT, not λ and T separately, and of two universal constants, C_1 and C_2. The variation of $e_{b\lambda}$ in W/m²µm with λ is shown in Figure 6.2.

The maximum value of $e_{b\lambda}$ occurs when λT is equal to 2898 µm-K or 5216 µm-°R. This can be seen when $e_{b\lambda}$ is plotted versus λ for different values of the temperature. Note that as the black body temperature decreases, the emitted radiation shifts to longer wavelengths. A good estimate of the relevant wavelength range can be obtained by using the function $F_{0-\lambda T}$. It is defined as the fraction of the total black body radiation that is emitted between a wavelength of 0 and a wavelength λ for a black body at temperature T,

$$F_{0-\lambda T} = \frac{\int_0^\lambda e_{b\lambda}(T)d\lambda}{\sigma T^4} \tag{6.6}$$

$F_{0-\lambda T}$ equals 0.5 when λT equals 4111 µm-K or 7400 µm-°R. One half of the black body energy is at wavelengths below this value ($\lambda = 4111/T$ µm), termed $\lambda_{1/2}$. Approximately 75% of the total black body energy occurs in the wavelength range between $0.5\lambda_{1/2}$ and $2\lambda_{1/2}$. Figure 6.3 illustrates how the energy spectrum varies with temperature. The wavelength interval that encompasses 80% of the black body energy at a given temperature is shown on the figure. At room temperature this spectrum ranges from the near infrared to the far infrared, while for solar energy at an effective black body temperature of 5780 K, the spectrum is centered on the visible and very near infrared.

Using the approximation given previously, for solar energy, $\lambda_{1/2}$ is 0.71 µm and 75% of the solar energy lies between 0.36 µm and 1.4 µm. For room temperature black bodies, $\lambda_{1/2}$ is 13.7 µm and 75% of the energy is between 6.9 and 27.4 µm.

This disparity in wavelength spectra gives rise to the atmospheric greenhouse effect. Earth's atmosphere is relatively transparent to incoming solar radiation in the

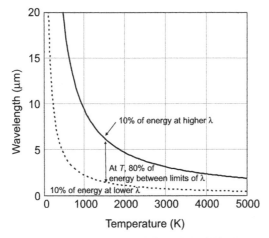

Figure 6.3. Wavelength interval containing 80% of black body emission.

visible and near IR bands. Some of the radiation emitted by the Earth's surface at longer infrared wavelengths is absorbed by gases in the atmosphere. Carbon dioxide is one of the important gases that absorb infrared radiation. As the CO_2 concentration increases, the atmosphere absorbs more of the radiation emitted at Earth's surface and some of this energy is reemitted back to the surface. The reemission to Earth impedes the net radiation from Earth's surface to space, which must ultimately balance the solar heating of the planet, and so the surface temperature rises in response.

Real bodies are not black, and they emit less energy than a black body at the same temperature. For real bodies, we define the emissivity as the ratio of the energy emitted from the surface of the real body to the energy emitted from the surface of a black body at the same temperature. The total emissivity, ε, is defined by comparing the energy emitted over all wavelengths, while the monochromatic emissivity, ε_λ, is defined by considering the energy emitted at a specific wavelength:

$$\varepsilon = \frac{e}{e_b} \quad \text{and} \quad \varepsilon_\lambda = \frac{e_\lambda}{e_{b\lambda}} \tag{6.7}$$

where e_λ is the monochromatic emissive power of the real body. For some materials ε_λ can vary substantially with wavelength. Figure 6.4 shows the behavior of ε_λ with wavelength for a particular ceramic.

For almost all most engineering situations, we can assume that ε_λ is equal to the monochromatic absorptivity α_λ at that wavelength. If the ceramic material shown in Figure 6.4 is at room temperature, then the energy emitted from the ceramic is centered in the mid to far infrared range. In this case, the total emissivity, which can be determined from

$$\varepsilon = \frac{\int_0^\infty \varepsilon_\lambda e_{b\lambda} d\lambda}{e_b} \approx \frac{\int_{5\mu m}^\infty \varepsilon_\lambda e_{b\lambda} d\lambda}{e_b} \tag{6.8}$$

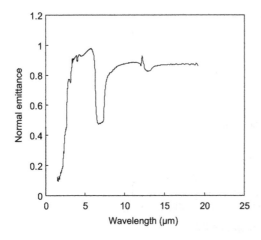

Figure 6.4. Monochromatic normal emittance (in the direction normal to a surface) of boron nitride at room temperature [6.1].

will have a value of about 0.8. If the incident radiation energy on the ceramic surface comes primarily from the sun, that energy is mainly in the visible and near infrared wavelengths. The total absorptivity of the surface is a function of both the surface properties and the properties of the incident radiation. The absorptivity of a body to incident solar radiation can be determined by

$$\alpha = \frac{\int_0^\infty \alpha_\lambda q_{i\lambda} d\lambda}{q_i} \approx \frac{\int_0^{3\mu m} \alpha_\lambda q_{i\lambda} d\lambda}{q_i} \tag{6.9}$$

where q_i is the incident radiation energy per unit time and area (i.e., flux), and $q_{i\lambda}$ is the energy per unit area and time on a particular wavelength. For the material shown on Figure 6.4, the total absorptivity α for incident solar radiation will have a low value, and in this case the total emissivity and total absorptivity of the ceramic are very different. On the other hand, if the source of incident energy striking the ceramic surface comes from a black body with a temperature close to the material's temperature, then the total absorptivity will have a value approximately the same as the total emissivity.

Assuming that $\varepsilon_\lambda \approx \alpha_\lambda$, there are several cases for which $\alpha \approx \varepsilon$ for a given body:

- ε_λ does not vary with λ over the range of wavelengths containing appreciable energy for both the emitted and incident radiation. In this case, the body is called a *gray* body.
- The body is a black body.
- The source of incident radiation is a black body at a temperature close to the receiving body.
- The source of incident radiation is a gray body at a temperature close to the receiving body.

In the special case where a small test body is placed inside the constant temperature enclosure shown in Figure 6.1, thermal equilibrium ensues when the test body reaches the temperature of the enclosure walls. *In this case*, it can be shown that for

Table 6.1. *Typical values of absorptivity for materials receiving radiation from a room temperature black body source and from solar energy*

Material	Black body source at 300 K	Solar
Aluminum foil bright	0.03	0.10
Aluminum weathered	0.2	0.54
Brick	0.90	0.63
Concrete rough	0.91	0.60
Paint white epoxy	0.85	0.25
Snow fresh	0.82	0.13

the test body $\varepsilon = \alpha$. This is one version of what is known as Kirchhoff's Law. Note that this is a special case covered by the aforementioned rules: the incident radiation on the test body is black body radiation at the enclosure temperature, and the test body and enclosure are at the same temperature. This *cannot* be extrapolated to conclude that in general ε will always equal α for the test body. For example, if the test body is a ceramic similar to Figure 6.4, $\varepsilon = \alpha$ when the test body is at equilibrium in the enclosure, but the body's emissivity will exceed its absorptivity value when the ceramic test body is at room temperature and is exposed to solar energy.

Simplified Approximation to Radiation Properties

Electromagnetic theory can be used to predict the radiation properties of pure homogeneous substances with optically smooth surfaces. However, it is far more difficult to predict the properties for a real material that may not be homogeneous and that may have a roughened surface with impurities such as oxide layers on the surface. Some estimated properties are given in the next section. Note that the absorptivity of a body is a function of the body's temperature as well as the temperature of the source of incident radiation. It must be stressed that these properties are only approximate. When more exact values are needed, it may be necessary to carry out detailed measurements. Even in that instance the properties may vary with time and exposure.

Some very general guidelines can be used to make a first estimate. For most engineering materials that are nonmetals, in the mid-to-far infrared range the monochromatic emissivity has a value near unity, typically between 0.7 and 1. Thus, at room temperature, nonmetals emit almost as much radiation as a black body at the same temperature. Similarly, they absorb a large fraction of the incident radiation emitted from other bodies near room temperature. For incident solar radiation, nonmetals may have widely different absorptivities. Because much of solar radiation is in the visible wavelengths, this can be judged by eye. Bodies that are white or light-colored will have a low solar absorptivity while bodies that appear dark to the eye will have high solar absorptivities. Table 6.1 gives some typical values. Note that the values given for solar absorptivity can vary widely for materials such as brick and concrete.

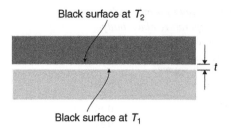

Black surface at T_2

Black surface at T_1

t Figure 6.5. Black bodies as part of built up wall.

For polished metals, the monochromatic emissivity tends to decrease at higher wavelengths. In the mid-to-far infrared range, a metal surface will have a much lower emissivity than a nonmetal and will absorb a small fraction of incident radiation from a source at room temperature. If an oxide layer builds up on the surface, the radiation properties of the surface will start to approach those of a nonmetal.

6.4 Heat Transfer Between Black Bodies, Linearized Formation

If two plane solid bodies are parallel and are separated by a small gap (Fig. 6.5), almost all of the radiation emitted from the surface of either body will be incident upon the second body. If the bodies are near room temperature and are nonmetals, they could, as a first-order approximation, be considered black bodies, giving an upper limit to the actual radiative exchange.

The radiation energy emitted by surface 1, when treated as a black body, is $A_1 \sigma T_1^4$. This energy is absorbed by surface 2, which is also treated as a black body. The radiant energy emitted by surface 2 is $A_2 \sigma T_2^4$ and is absorbed by surface 1. Each of these represents an energy transfer in a single direction. The net radiant energy between the two surfaces becomes

$$Q_R = A_1 \sigma \left(T_1^4 - T_2^4 \right) \tag{6.10}$$

For an air-filled gap, the radiation acts in parallel with conduction or convection through the air. When the thickness of the gap, t, is small, convection is suppressed and heat transfer through the air is essentially by conduction. The net heat transfer becomes

$$Q_{net} = \frac{k_{gas} A_1}{t} (T_1 - T_2) + A_1 \sigma \left(T_1^4 - T_2^4 \right) \tag{6.11}$$

When the two surfaces are part of a more complex wall geometry containing multiple layers the solution is not so straightforward. The two temperatures in Eq. (6.11) are unknown and the equation is nonlinear. The calculation can be simplified by linearizing the radiation term around the mean temperature between T_1 and T_2. If the mean temperature is not known at the outset, we also need to inquire how sensitive the linearized form is to the value assumed for the mean temperature. To linearize the radiation flux, the two temperatures will be written in terms of the

Table 6.2. Values of h_r, radiation heat
transfer coefficient

T_M (°C)	T_M (K)	h_r (W/m² K)
20	293	5.7
100	373	11.8
500	773	107
800	1073	280

mean temperature,

$$T_m = \frac{T_1 + T_2}{2} \text{ and } \Delta T = T_1 - T_2 \tag{6.12}$$

$$T_1 = T_M + \frac{\Delta T}{2} \tag{6.13}$$

$$T_2 = T_M - \frac{\Delta T}{2} \tag{6.14}$$

Expanding Eq. (6.10) and substituting these two expressions, we have

$$Q_R = \sigma A_1 (T_1^2 + T_2^2)(T_1 + T_2)(T_1 - T_2) \tag{6.15}$$

$$Q_R = \sigma A_1 \left(T_M^2 + T_M \Delta T + \frac{(\Delta T)^2}{4} + T_M^2 - T_M \Delta T + \frac{(\Delta T)^2}{4} \right)(2T_M)(T_1 - T_2) \tag{6.16}$$

$$Q_R = \sigma A_1 \left(2T_M^2 + 2\frac{(\Delta T)^2}{4} \right)(2T_M)(T_1 - T_2) \tag{6.17}$$

When the term involving the temperature difference, $(\Delta T)^2$, is small compared to T_M^2 (remember the temperatures must be given in an absolute scale), the first bracket on the right-hand side of Eq. (6.17), can be simplified, and the equation becomes

$$Q_R = \sigma A_1 \left(4T_M^3 \right)(T_1 - T_2) \tag{6.18}$$

when $(\Delta T)^2/T_M^2 \ll 1$.

We may now define a radiation heat transfer coefficient, h_r, for black body radiation as

$$h_r = 4\sigma T_M^3 \tag{6.19}$$

$$Q_R = h_r A_1 (T_1 - T_2) \tag{6.20}$$

Figure 6.6 shows the error associated in using h_r instead of the exact formulation given in Eq. (6.10). For ratios of T_2/T_1 less than 1.3, the error is 2% or less.

Table 6.2 presents the magnitude of h_r for different mean temperatures. These values can be compared to the table of values given in Chapter 2 for different modes of convection heat transfer. Near room temperature the value of the h_r is the same

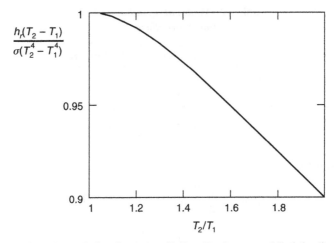

Figure 6.6. Linearized formulation for net radiation flux between black bodies compared to the exact value.

magnitude as the heat transfer coefficient for natural convection in air. At higher temperature levels, radiation becomes equivalent to forced convection; and in some applications, such as furnaces, radiation may be the dominant mode of heat transfer.

When properly applied, the error associated with the linearization is usually small; however, there is another important consideration. When the temperatures of the radiating bodies are not initially known, the value of the mean temperature, T_M, is uncertain. This is a more critical concern. Figure 6.7 illustrates how errors in the value of T_M influence the value of h_r. To maintain h_r accuracy to $\pm 10\%$, T_M cannot vary by more than $\pm 3\%$ of its absolute temperature value. For bodies near room temperature, an error of $\pm 10°C$ in T_M results in an error of $\pm 10\%$ in h_r.

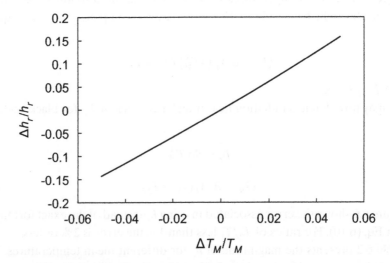

Figure 6.7. Error in linearized radiation coefficient $\Delta h_r / h_r$ as a function of error in the mean temperature $\Delta T_M / T_M$.

Figure 6.8. Plane and solid angles.

6.5 Geometry and Configuration Factors

Because radiation acts at a distance, all surrounding sources of radiation that can interact with a given body must be considered. The geometry of these interactions can lead to formidable mathematical problems. Fortunately, a number of simplifying techniques can be used in engineering evaluations.

The concept of solid angles will be used to deal with the radiation in three-dimensional space. In plane geometry, an angle is defined by the ratio of the inscribed linear arc to the radius from the origin:

$$\theta = s/r \tag{6.21}$$

In three-dimensional geometry, the solid angle is defined as the ratio of the included area of a segment of a spherical surface to the square of the radius from the origin (Fig. 6.8):

$$\omega = A/r^2 \tag{6.22}$$

For the solid angle, the area A represents the included area on the spherical surface surrounding the origin, not the projected plane area. For a hemisphere, the solid angle about the origin is 2π, while for a sphere the solid angle is 4π.

Returning to the enclosed space kept at constant temperature shown in Figure 6.1, a surface positioned anywhere within the space would have black body radiation falling on it. An imaginary or transparent surface such as that shown in Figure 6.9 within the enclosed space would have black body radiation passing through it. To an observer above, the surface the radiation could equally appear to be emitted from the surface acting as a black body.

Adopting the particle view of radiation, consider the photons passing through a small area dA normal to the photon trajectory. In any direction, an equal number of photons would pass per unit time through a surface subtended by a given solid angle regardless of the distance from the emitting surface (Fig. 6.9). If we think of a small solid angle $d\omega$ around the direction normal to the surface, the number of photons passing, a measure of a one-way energy flux, would be proportional to

$$dQ_{\text{oneway}} \sim dA d\omega \tag{6.23}$$

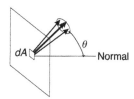

Figure 6.9. One-way radiant energy flux.

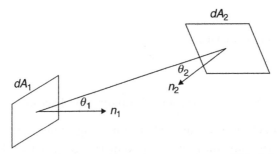

Figure 6.10. Radiation exchange between two small black bodies.

If instead, we think of photons passing through the surface at an angle θ to the normal, the one-way energy flux would be proportional to the projected area of the surface normal to the propagation direction, $\cos\theta\, dA$:

$$dQ_{\text{oneway}} \sim dA\cos\theta\, d\omega \qquad (6.24)$$

Intensity of Radiation

The intensity of radiation, either emitted from the surface or passing through an imaginary surface in one direction within a small solid angle, can be defined as

$$i = \frac{\text{energy within solid angle in direction } S}{(\text{time})\,(\text{solid angle})\,(\text{area normal to } S)} \qquad (6.25)$$

If the radiation streaming in one direction is not attenuated or augmented the intensity remains constant within $d\omega$ irrespective of the distance from the surface. The one-way energy flow per unit time becomes

$$dQ_{\text{oneway}} = (i\, d\omega)(\cos\theta\, dA) \qquad (6.26)$$

Consider two small black body surfaces, shown in Figure 6.10, with areas dA_1 and dA_2 and temperatures T_1 and T_2, respectively.

The amount of radiant energy emitted by black body 1 that is directed to black body 2 is given by

$$dQ_{1\to2} = i_{b1}(\theta_1)dA_1\cos\theta_1\, d\omega_{12} = \frac{i_{b1}(\theta_1)dA_1\cos\theta_1\, dA_2\cos\theta_2}{r^2} \qquad (6.27)$$

where i_b represents the intensity of radiation from a black body. Similarly, the amount of radiation from surface 2 to surface 1 is given by

$$dQ_{2\to1} = i_{b2}(\theta_2)dA_2\cos\theta_2\, d\omega_{21} = \frac{i_{b2}(\theta_2)dA_2\cos\theta_2\, dA_1\cos\theta_1}{r^2} \qquad (6.28)$$

Because the surfaces are black bodies, all of the radiation directed toward them is absorbed and the net radiant transfer between the two surfaces is given by the difference in the heat transfers expressed in Eqs. (6.27) and (6.28). If the two black bodies are at the same temperature, by the second law of thermodynamics, the net heat transfer must be zero and Eqs. (6.27) and (6.28) must be equal. Thus, for any black bodies at the same temperature, the intensity of emitted radiation is the same, and, moreover, for a black body the intensity of radiation is the same for any angle:

$$i_b(\theta_1) = i_b(\theta_2) \tag{6.29}$$

This conclusion applies only to black bodies.

Consider the situation when surface A_2 forms a hemisphere centered over differential surface dA_1, a black body. The one-way radiative flux from dA_1 to A_2 is equivalent to the black body emissive power of dA_1:

$$dQ_{1\rightarrow 2} = i_{b1}dA_1 \int_{2\pi} \cos\theta_1 d\omega_{12} = dA_1 \sigma T_1^4 \tag{6.30}$$

The integral in Eq. (6.30) can be shown to yield π. Thus the black body intensity can be related to the emissive power as

$$i_{b1} = \frac{\sigma T_1^4}{\pi} \tag{6.31}$$

If the two black bodies A_1 and A_2 are finite in extent and each is at a uniform temperature the one-way flux from surface 1 to surface 2 can be found by integrating Eq. (6.27) over the two surface areas,

$$Q_{1\rightarrow 2} = \int_{A2} \int_{A1} \frac{i_{b1}(\theta_1) \cos\theta_1 \cos\theta_2 dA_1 dA_2}{r^2} \tag{6.32}$$

The surface A_1 is isothermal so that the intensity does not vary with location over A_1. We have shown above that for a black body the intensity doesn't vary with angle. With these conditions the intensity i_{b1} is a constant and can be removed from within the integral. This provides a substantial simplification because the integral now involves only geometry between A_1 and A_2:

$$Q_{1\rightarrow 2} = i_{b1} \int_{A2} \int_{A1} \frac{\cos\theta_1 \cos\theta_2 dA_1 dA_2}{r^2} \tag{6.33}$$

Using the relationship between black body intensity and emissive power given by Eq. (6.31), Eq. (6.33) can be rewritten as

$$Q_{1\rightarrow 2} = A_1 \sigma T_1^4 \frac{1}{A_1} \int_{A2} \int_{A1} \frac{\cos\theta_1 \cos\theta_2 dA_1 dA_2}{\pi r^2} = A_1 \sigma T_1^4 F_{12} \tag{6.34}$$

$$F_{12} = \frac{1}{A_1} \int_{A2} \int_{A1} \frac{\cos\theta_1 \cos\theta_2 dA_1 dA_2}{\pi r^2} \tag{6.35}$$

The quantity $A_1 \sigma T_1^4$ represents the amount of energy emitted by black body of area A_1 in all directions above its surface. The remaining terms in Eq. (6.34), rewritten as F_{12} in Eq. (6.35), represent the fraction of the black body energy emitted from surface A_1 that is incident upon surface A_2. F_{12} is known as the configuration factor or view factor. Because we were able to remove the intensity from the integral in Eq. (6.32),

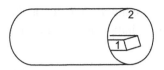

Figure 6.11. Tube furnace.

the configuration factor depends only on the geometry and not on the temperatures of A_1 and A_2.

6.6 The Configuration Factor

The radiation transfer between isothermal black bodies can be determined when the configuration factor is known. Evaluation of the integral in Eq. (6.35) can be formidable, so it is best to first seek known solutions and approximations to the configuration factor before resorting to lengthy "brute force" methods. Some simple relationships can be established. Because the configuration factor represents the fraction of energy emitted from a black body surface A_1 that is incident upon another surface, when all the radiation from surface A_1 is directed at a total of N surfaces (including back to A_1 itself if it is concave), then

$$\sum_{j=1}^{N} F_{1j} = 1 \qquad (6.36)$$

In words, 100% of the radiation from A_1 reaches the N surfaces present, each receiving a fraction F_{1j}. Following logic similar to the development of Eq. (6.34), the fraction of black body energy emitted from surface 2 incident upon surface 1 is

$$Q_{2\to1} = A_2 \sigma T_2^4 \frac{1}{A_2} \int_{A2} \int_{A1} \frac{\cos\theta_1 \cos\theta_2 dA_1 dA_2}{\pi r^2} = A_2 \sigma T_2^4 F_{21} \qquad (6.37)$$

When black bodies 1 and 2 are at the same temperature, the net radiative heat transfer between them must be zero. Therefore, Eqs. (6.34) and (6.37) are identical when the temperatures are equal and the following geometrical reciprocity relationship can be stated

$$A_1 F_{12} = A_2 F_{21} \qquad (6.38)$$

This relationship holds for any three-dimensional shape of the two surfaces, and it does not require the surface temperatures to be the same. It also holds when the view from surface 1 to surface 2 is partially obscured by a third surface.

Example 6.1 *Laboratory tube furnace* A simple laboratory furnace is made by wrapping electrical heater wire around the exterior of a cylindrical ceramic tube. Radiation from the inside cylindrical surface will be transmitted to an object within the furnace as well as back to the tube walls. In the case shown in Figure 6.11, a bar with rectangular cross-section is inserted in the furnace. We need to calculate the configuration factor from the inside surface of the tube to itself as well as to the bar within.

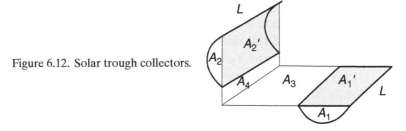

Figure 6.12. Solar trough collectors.

Surface A_2 represents the entire inside surface of the cylinder and A_1 represents the total surface area of the bar. To obtain an upper limit for the configuration factor from surface 2 to itself, F_{22}, assume that the cylinder length is much greater than its diameter. In this instance all of the radiation from surface 2 is directed toward surface 1 or toward itself:

$$F_{22} + F_{21} = 1 \qquad (6.39)$$

Surface 1 is convex so all of its radiation goes to surface 2 and F_{12} is unity if we neglect radiation to the open ends. Using the reciprocity relationship,

$$F_{21} = \frac{A_1 F_{12}}{A_2} = \frac{A_1}{A_2} \qquad (6.40)$$

Using Eq. (6.39),

$$F_{22} = 1 - F_{21} = 1 - \frac{A_1}{A_2} \qquad (6.41)$$

Note that in this case the location of the bar within the cylinder does not influence the result for F_{22}. For cases when the inside cylinder surface is not a black body, the location of the bar will influence the distribution of net heat transfer from the surface of the cylinder.

Numerous authors have calculated configuration factors for a number of different geometrical arrangements (see, for example, Howell [6.2]). The results are given as equations and graphs in a number of books devoted to radiation heat transfer, some of which are listed at the end of this chapter. Appendix B lists an abbreviated set of relationships for configuration factors between two bodies of finite sizes. These results can be extended to a much wider array of geometry by simple algebraic relationships and simplifications. The following two examples illustrate this.

Example 6.2 *Solar trough collectors* Two solar thermal collector panels are half cylinders. One panel is horizontal and the second is upright facing South, as shown in Figure 6.12. The cylindrical surfaces are A_1 and A_2. We need to find the radiative transfer beween the two panels.

The first task is to determine the configuration factor F_{12}. The configuration factor of the two cylindrical surfaces is definitely not contained in readily available tabulated relationships for configuration factors.

First, we must relate the configuration factors between the concave cylindrical surfaces to those between planar surfaces $A_{1'}$ and $A_{2'}$ that cover the mouth of the two cylindrical surfaces. The next step will relate the configuration factors between $A_{1'}$ and $A_{2'}$ to those given in the standard tables. The fraction of energy emitted by surface 1 that is directed to surface A_2 must pass through surface $A_{2'}$, so examining the latter is sufficient. Combining this with the reciprocity relationship, we find

$$F_{12} = F_{12'} = F_{2'1} \frac{A_{2'}}{A_1} \tag{6.42}$$

All the radiation that would be emitted by the imaginary surface $A_{2'}$ toward surface A_1 must pass through surface $A_{1'}$, so that $F_{2'1}$ is equal to $F_{2'1'}$. Again using reciprocity, F_{12} can be expressed as follows:

$$F_{12} = F_{2'1'} \frac{A_{2'}}{A_1} = F_{1'2'} \frac{A_{1'}}{A_1} \tag{6.43}$$

We have obtained an expression for F_{12} in terms of configuration factors between the two planar surfaces $A_{1'}$ and $A_{2'}$. However, the tabulated values of configuration factors give values only when the two surfaces are at right angles and are joined along a common side. We need to express $F_{1'2'}$ in terms of known configuration factors. To do that we will add imaginary areas A_3 and A_4. Then the total radiative energy flux from the combined areas 1' and 3 to the combined areas 2' and 4 is given by

$$(A_{1'} + A_3)F_{(1'+3)(2'+4)} = A_3(F_{34} + F_{32'}) + A_{1'}(F_{1'4} + F_{1'2'}) \tag{6.44}$$

In this expression, $F_{(1'+3)(2'+4)}$ as well as F_{34} can be found in tables. Note that $F_{(1'+3)(2'+4)}$ is not simply the sum of $F_{34}, F_{32'}, F_{1'4}$ and $F_{1'2'}$ since the areas of the surfaces differ. The value of $F_{1'2'}$ can be determined from Eq. (6.44) when $F_{32'}$ and $F_{1'4}$ are known. To find configuration factor $F_{32'}$, we use the expression

$$F_{32'} = F_{3(2'+4)} - F_{34} \tag{6.45}$$

and similarly

$$F_{1'4} = \frac{A_4}{A_{1'}} F_{41'} = \frac{A_4}{A_{1'}} [F_{4(31')} - F_{43}] \tag{6.46}$$

The value of F_{12}, our initial objective, can now be found by combining Eqs. (6.43) through (6.46).

Example 6.3 *Approximating a laboratory black body* For surfaces that are not black, the overall emission will approach that of a black body if the surface contains many cavities. One example is a plane surface of area A_3 that contains evenly spaced drilled holes as shown on Figure 6.13. A_3 is the area of the plate before the holes are drilled. The holes have a cylindrical portion, of area A_1 (this is the total area for all the cylindrical surfaces within the holes) and a conical section A_2 at the bottom. If the surfaces of the holes are not black bodies, the

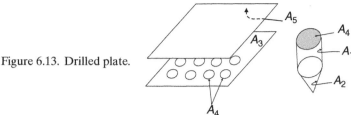

Figure 6.13. Drilled plate.

solution for the net radiation emitted from the holes to the exterior will require the configuration factors between the surfaces of the hole and the exterior.

As a first step we will estimate the configuration factor from A_2 and from A_1 to a plane surface A_5 parallel to and of equal area to A_3. We want to find the configuration factors from surfaces A_1 and A_2 to surfaces above plate A_3.

Using reciprocity,

$$A_2 F_{25} = A_5 F_{52} \quad \text{and} \quad A_1 F_{15} = A_5 F_{51} \tag{6.47}$$

Next, we may construct an imaginary plane surface A_4 at the mouth of the holes, since the holes are evenly spaced,

$$F_{54} = F_{53} \frac{A_4}{A_3} = F_{51} + F_{52} \tag{6.48}$$

and F_{53} can be found from standard tables of configuration factors.

The upper limit for F_{25} can be found by assuming that all the radiation from A_2 that passes through A_4, represented by F_{24}, is directed toward A_5. F_{25} can be found by evaluating F_{42} (see the homework for an example). The result is an upper limit for F_{25}:

$$F_{25\,\mathrm{Max}} = F_{24} = \frac{A_4 F_{42}}{A_2} \tag{6.49}$$

Combining this with Eq. (6.48) will give an estimate of F_{51} and F_{15}. This will be a lower limit for F_{15}. Note that a poor estimate of F_{25} is obtained from the expression

$$F_{25} \neq F_{24} F_{45} \tag{6.50}$$

This product is equivalent physically to taking the radiation emitted from A_2 directed toward to A_4, which is concentrated in angles close to the normal to A_4 as well as the normal to A_5, and spreading it out so that it has equal intensity in all directions as it leaves A_4.

A lower limit for F_{25} can be found by assuming F_{15} equals F_{14}. This yields an upper limit for F_{15}, by using Eq. (6.48) to find F_{25}. These results will be used in a later section to show how the plate with the drilled holes approaches black body behavior.

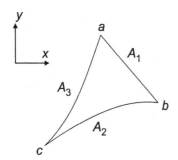

Figure 6.14. Extruded cross section with $z \gg x, y$.

Two-Dimensional Approximations

To be comprehensive, we must consider all of the bodies that exchange radiation with a given body. We can construct a closed surface made up of real or imaginary surfaces that surround the body in question. When one of the orthogonal dimensions is much larger than the other two for all of the bodies exchanging radiation, a relatively simple method of obtaining configuration factors is available. In that instance, radiation to the end planes of the closed surface, normal to the longest dimension, can be neglected. First, consider the special case of three plane or convex surfaces that radiate to each other and form a closed volume. Because they are convex they do not radiate to themselves. Neglecting the radiation to the end planes, we can write for each surface, using reciprocity,

$$A_1 = A_1 F_{12} + A_1 F_{13} \tag{6.51}$$

$$A_2 = A_2 F_{21} + A_2 F_{23} = A_1 F_{12} + A_2 F_{23} \tag{6.52}$$

$$A_3 = A_3 F_{31} + A_3 F_{32} = A_1 F_{13} + A_2 F_{23} \tag{6.53}$$

If we add the first two equations and subtract the third all of the configuration factors cancel except F_{12}. The resulting expression is

$$F_{12} = \frac{A_1 + A_2 - A_3}{2A_1} \tag{6.54}$$

If the three surfaces form an extruded cross section that remains unchanged at different values of z the coordinate normal to the cross section shown in Figure 6.14, the areas can be related to the length of the surfaces in the x-y plane. A_1 is $z(ab)$, A_2 is $z(bc)$ and A_3 is $z(ac)$.

Equation (6.50) can now be written in terms of the length of the line segments:

$$F_{12} = \frac{ab + bc - ac}{2ab} = \frac{\sum \text{Length of both Surfaces in } F_{12} - \text{Length of Third Surface}}{2 \,(\text{Length of Emitting Surface})}$$

$$\tag{6.55}$$

This relation can be used in a more general extruded cross section involving more than three surfaces and allowing concave as well as convex surfaces (Fig. 6.15). As

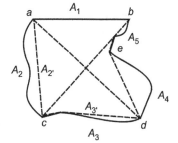

Figure 6.15. General extruded cross section.

before, the surface dimensions in the x-y plane most be smaller than the z dimension so that radiation to the ends can be neglected.

In this case, assume we are looking for the configuration factor between A_1 and A_3. The imaginary lines connecting all of the end points of A_1 and A_3 have been constructed. In addition the surface $A_{3'}$ has been added so that it is convex or planar everywhere. Just as a previous example, F_{13} is equal to $F_{13'}$. The imaginary lines can be visualized as strings connecting the respective points. All of the strings are "pulled tight" so that they all form convex or planar surfaces. Now the desired configuration factor F_{13} can be found from the following expression:

$$F_{13} = F_{13'} = 1 - F_{12'} - F_{1(4'5')} \tag{6.56}$$

where $F_{12'}$ is the configuration factor from A_1 to the imaginary planar surface formed by the string (ac) and $F_{1(4'5')}$ is the configuration factor from A_1 to the imaginary convex surface formed by string (deb). Note that the strings ab, ac, and bc form a three-body enclosure so that

$$F_{12'} = \frac{ab + ac - bc}{2ab} \tag{6.57}$$

and that ab, ad, and deb form another three-body enclosure so that

$$F_{1(4'5')} = \frac{ab + deb - ad}{2ab} \tag{6.58}$$

Combining these to relationships with Eq. (6.56),

$$F_{13} = \frac{2ab - (ab + ac - bc) - (ab + deb - ad)}{2ab} = \frac{bc + ad - ac - deb}{2ab} \tag{6.59}$$

This expression can be rewritten by remembering that strings are attached to the endpoints of the two surfaces involved in the configuration factor; then

$$F_{13} = \frac{\sum \text{Length of Crossed Strings} - \sum \text{Length of Uncrossed Strings}}{2 \,(\text{Length of } A_1)} \tag{6.60}$$

This method, referred to as the crossed-strings technique, was originally developed for lighting calculations before being carried over to radiation heat transfer.

Example 6.4 *Use of crossed-strings method.* Figure 6.16 illustrates a cross section of a new façade system being used on commercial buildings, a double skin

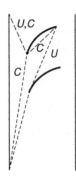

Figure 6.16. A double skin façade.

façade. A sun shading system of curved louvers is placed between two sheets of glass. Ambient temperature air is circulated in the cavity to cool the louvers in summer. In the infrared wavelengths, the glass approaches black body behavior. We need to determine the configuration factor between the glass and the upper and lower surfaces on the louvers to estimate the radiant heat transfer through the façade.

The horizontal length of the louvers is much larger than their width and the spacing of the cavity. The cross section can be assumed to be extruded with a negligible view factor to the end walls. The figure shows the construction of the crossed and uncrossed strings, marked C and U respectively, between one glass sheet and the lower surface of one louver. In situations such as this the identification of crossed and uncrossed may not be obvious. One technique to visualize these is to rotate or displace one object so that the crossed and uncrossed strings are not overlapping.

Limiting Case

The use of the crossed-strings approximation assumes that the dimension normal to the cross section, z in Figure 6.14, is large. Figure 6.17 gives some measure of the level of approximation. In this case, two parallel rectangular plates are considered where the plate width, x, is equal to the plate separation. The exact solution for the view factor (obtained numerically) is compared to the view factor derived by assuming the plate length, the z dimension, is very large, so that the crossed-string approximation

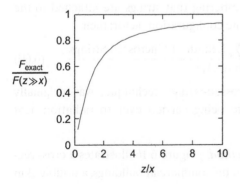

Figure 6.17. Exact versus crossed strings configuration factors for parallel rectangular plates. Plate width x, plate length z, and spacing x.

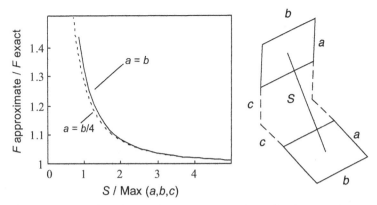

Figure 6.18. Comparison of approximate values, based on Eq. (6.62), to exact values for configuration factors for surfaces with orthogonal normal vectors.

can be used. It can be seen in this instance when z/x is about 8 or larger, the error in using the crossed strings is less than 10%.

Numerical Approximation of Configuration Factors

Several analytical methods can be used to evaluate the configuration factor. In some instances, the integral of Eq. (6.35) must be evaluated numerically. The "brute force" approach involves dividing A_1 and A_2 into many smaller subareas, ΔA_i, ΔA_j respectively, and summing them,

$$F_{12} = \frac{1}{A_1} \int_{A2} \int_{A1} \frac{\cos\theta_1 \cos\theta_2 dA_1 dA_2}{\pi r^2} \approx \frac{1}{A_1} \sum_{A_2} \sum_{A_1} \frac{\cos\theta_i \cos\theta_j \Delta A_i \Delta A_j}{\pi r_{ij}^2} \quad (6.61)$$

For this summation, r_{ij} represents the distance from the center of element ΔA_i to the center of element ΔA_j and the angles are measured between each element's normal and r_{ij}. Figures 6.18 can be used to judge how small the subarea needs to be for accurate results using the summation on the right-hand side of Eq. (6.61). Generally, when the center-to-center distance between elements is 2.5 to 3 times the largest linear dimension of ΔA_i and ΔA_j, the error in the approximation is less than 15%. Note that this means that fairly large surface elements can be evaluated without subdivision when they are still reasonably close to each other, that is,

$$F_{12} \approx \frac{1}{A_1} \frac{\cos\theta_1 \cos\theta_2 A_1 A_2}{2 r_{12}^2} \quad \text{when} \quad r_{12} \geq 3(\text{Max of } X_1, Y_1, X_2, Y_2) \quad (6.62)$$

Students frequently try to estimate the configuration factor between two finite surfaces as the solid angle measured from the center of one surface to the second surface. This will be inaccurate especially when the surface normals are orthogonal and the term $\cos\theta_1$ in Eq. (6.62) becomes much less than unity.

Monte Carlo methods may also be used to numerically evaluate configuration factors. Before attempting to set these up these, it might be less time consuming to survey commercial numerical heat transfer programs.

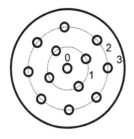

Figure 6.19. Fuel rod configuration.

6.7 Radiation Heat Transfer Between Black Bodies

With the radiation geometry now in hand by use of configuration factors, we can deal with radiant heat transfer between different bodies. In these cases, the medium separating the bodies, usually air, will be assumed to be transparent. There are cases in which this assumption does not hold, for example, in large combustion chambers with high concentrations of water vapor, carbon dioxide, and carbon particles and in some instances, large interior volumes of buildings with reasonable high humidity levels.

Consider the example of a transport cask for spent nuclear fuel rods. Figure 6.19 shows a cross section with the rods arranged in concentric rings within a cylindrical container. The rods may still produce residual heating, and limiting the temperature of the rods during transport is important for safety. We will assume the rods in each concentric ring have the same temperature and internal heat generation rate. In addition, the surface of each rod will be assumed to be at uniform temperature.

For this example, the surfaces of the rods are likely to be oxidized, so that they approach black body behavior at moderate temperatures. All of the rods in one ring will be grouped together with a single area equal to the sum of their surface area. The length of the rods and the cylindrical case is much larger than the diameter, so for simplicity the radiation to the end caps of the cylinder will be ignored.

For the rods in ring 1, the net radiation emitted from their surfaces to themselves and other bodies is

$$Q_{1\text{Emitted}} = A_1 \sigma T_1^4 = A_1 F_{10} \sigma T_1^4 + A_1 F_{11} \sigma T_1^4 + A_1 F_{12} \sigma T_1^4 + A_1 F_{13} \sigma T_1^4 \qquad (6.63)$$

The radiation emitted from the other rods and the inside of the cylinder that is absorbed by A_1 is

$$Q_{1\text{Absorbed}} = A_0 F_{01} \sigma T_0^4 + A_1 F_{11} \sigma T_1^4 + A_2 F_{21} \sigma T_2^4 + A_3 F_{31} \sigma T_3^4 \qquad (6.64)$$

The net radiation leaving surface A_1 becomes, using reciprocity,

$$Q_{1\text{Net}} = A_1 F_{10}(\sigma T_1^4 - \sigma T_0^4) + A_1 F_{12}(\sigma T_1^4 - \sigma T_2^4) + A_1 F_{13}(\sigma T_1^4 - \sigma T_3^4) \qquad (6.65a)$$

Alternatively, written in linearized form, the expression becomes

$$Q_{1\text{Net}} = A_1 F_{10} h_r (T_1 - T_0) + A_1 F_{12} h_r (T_1 - T_2) + A_1 F_{13} h_r (T_1 - T_3) \qquad (6.65b)$$

The physical behavior embedded in Eq. (6.65a) can be represented by an electrical analogy, where Q is analogous to current flow, emissive power to potential, and

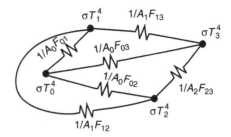

Figure 6.20. Electrical analogy for black bodies.

$1/A_iF_{ij}$ to resistance. This is shown on Figure 6.20. In this form, the electrical analogy cannot be directly used with the electrical analogy for conduction heat transfer: the radiative version uses the emissive power as the potential while the conductive version uses temperature.

To use the electrical analogies for conduction and radiation together, the linearized form of the radiation expression for the net radiation, Eq. (6.65b), must be applied instead. Figure 6.21 shows this linearized electrical analogy. If there are significant temperature differences between the elements, the linearized radiation term, h_r, from Eq. (6.19) must be evaluated separately for each pair of objects based on their respective mean temperature (while also being aware of the error limits illustrated in Fig. 6.7). When all of the surfaces are black, the linearized form of the radiation transfer can be coupled with any convection or conduction and internal heating to carry out an overall energy balance for the system.

Note that with four separate surface areas (nodes 0 through 3), the electrical analogy is becoming quite complicated. One important reason for using the graphical representation of the heat transfer is to help visualize the process. When the number of nodes exceeds four, in general the electrical analogy becomes too complicated to be of much help although there are special cases where it still may be of assistance.

One simplified case would occur if the number of rods in each concentric ring was increased until neighboring rods almost touched. In that case the rods in each concentric ring can exchange radiation with the rods only in the nearest neighboring rings. The electrical analogy for this case is shown in Figure 6.22.

The electrical analogy can be simplified by omitting the area of one element when it is much smaller than the other element if they all have comparable emissive powers. On the other hand, if a surface is adiabatic, it may still serve as an important

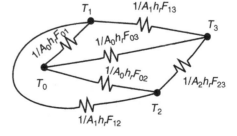

Figure 6.21. Linearized electrical analogy.

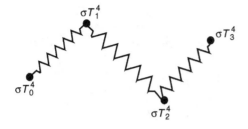

Figure 6.22. Rods touching in each concentric ring.

intermediate path between two other elements: it absorbs radiation from one surface and reemits to the surfaces surrounding it.

   If the surface temperature of a black body object is uniform, the simple relationships given in the preceding text will hold; but when the temperature varies along the surface of a black body, further care is required. Remember that the incident radiation to a body is an integration of the radiation emitted over the entire surface of a second body. If the latter has a nonuniform temperature, it may still be adequate to represent it as isothermal with a temperature that is found from the mean of T^4 over the surface. This approximation is more applicable when the dimensions of the second surface are smaller than the distance between the two bodies and the view factor from first body to different subareas of the second body doesn't vary substantially. Upper and lower limits of this approximation can be obtained by treating the body as isothermal at its extreme upper and lower surface temperatures. It may be that the contribution from the nonisothermal surface to the total radiation flux is modest. A more detailed check of the isothermal approximation can be made by dividing the second surface into several subareas, each with its own mean temperature.

6.8 Radiation Heat Transfer Between Nonblack Bodies

In many cases the bodies exchanging radiation are not black. In the most general situation the emitted radiation varies with orientation from the surface normal as well as location along the surface. The latter variation may be due to different temperatures along the surface or varying emissivity, for example, due to different rates of oxidation. The reflected radiation may be a function of the orientation of the incoming flux relative to the surface normal as well as the varying surface finish and chemistry.

   In keeping with the spirit of this text, we seek to find a simplified model for radiant transfer for nonblack surfaces. If the two bodies A_1 and A_2 are finite in extent, the one-way flux leaving surface 1 that is *directed toward* (but not necessarily absorbed by) surface 2 can be found by integrating Eq. (6.27) over the two surface areas:

$$Q_{1 \to 2} = \int\limits_{A_2} \int\limits_{A_1} \frac{i_1(\theta_1) \cos\theta_1 \cos\theta_2 \, dA_1 \, dA_2}{r^2} \tag{6.66}$$

   In this case $i_1(\theta_1)$ inside the integral represents the intensity of the combined emitted and reflected radiation leaving surface A_1 at a particular location, say x_1, y_1, in the θ_1 direction. If $i_1(\theta_1)$ can be considered a constant for all angles θ_1 and all locations on A_1, it can be removed from the integral. If this is possible, then the remaining integral can be directly related to the configuration factors.

The intensity $i_1(\theta_1)$ of the radiation leaving surface A_1 is made up of two parts: the radiation emitted from surface A_1 and the radiation reflected from A_1. So, $i_1(\theta_1)$ can be represented as

$$i_{1\text{Leaving}}(\theta_1) = i_{1\text{Emitted}}(\theta_1) + i_{1\text{Reflected}}(\theta_1) \tag{6.67}$$

The emitted portion will remain constant if the following three conditions are met.

1. The intensity of emitted radiation does not vary with angle.

Many common surfaces exhibit this trait except at values of θ approaching 90 degrees. These are referred to as diffuse emitters. The total radiation emitted from the surface per unit area is $\varepsilon_1 \sigma T_1^4$ and the intensity of emitted radiation from a diffuse emitter is $\varepsilon_1 \sigma T_1^4/\pi$.

To ensure that the intensity of the emitted radiation does not vary with location along A_1, the following must hold:

2. The temperature is uniform over surface A_1.
3. The emissivity is uniform over A_1.

If the temperature of all the surfaces exchanging radiation is of the same magnitude and the monochromatic emissivity does not vary significantly over the wavelengths containing most of the radiant energy, then the total absorptivity and emissivity will be roughly the same. In addition, if the bodies are opaque, the reflectivity ρ becomes equal to $1 - \alpha$ or $1 - \varepsilon$. For the reflected portion of the intensity in Eq. (6.67) to be a constant, the following additional conditions must hold:

4. The body is a diffuse reflector, so that the intensity of reflected energy does not vary with angle.
5. The magnitude of the radiant energy per unit area incident on surface A_1 is uniform over the surface.

For most real nonblack surfaces, it is rare that these five conditions are met exactly. We will consider some approximate cases. When the five conditions hold, the intensity leaving the surface can be represented as

$$i_{1\text{Leaving}} = \frac{1}{\pi}\left[\varepsilon_1 \sigma T_1^4 + \left(\frac{1-\varepsilon_1}{A_1}\right)\sum_j Q_{j\to 1}\right] \tag{6.68}$$

where the last term represents all of the radiation incident on A_1. Under these conditions, $i_{1\text{Leaving}}$ is constant and can be removed from the integral in Eq. (6.66). It follows from the definition of configuration factor that the radiant energy leaving A_1

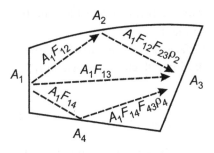

Figure 6.23. One bounce radiation heat transfer.

directed toward A_2 can be written as

$$Q_{1\to2} = \left[\varepsilon_1 \sigma T_1^4 + \frac{(1-\varepsilon_1)}{A_1} \sum_j Q_{j\to1} \right] A_1 F_{12} \tag{6.69}$$

A similar expression holds for every surface that participates in the radiative exchange. We will see later how to obtain an exact solution for this set of equations.

Radiant Exchange between Bodies with a High Emissivity: One-Bounce Approximation

For bodies at moderate temperature, all of the radiation occurs in the mid to far infrared wavelengths. Many common materials, except for polished metals, have a high emissivity in these wavelengths. If all the bodies exchanging radiation are at about the same temperature level, then for each surface the emitted portion of the intensity will be substantially larger than the reflected portion. In this case, conditions 4 and 5 above do not have to be strictly followed to allow Eq. (6.69) to hold. Note that the incident radiation on the surface of a body is a function of the radiation leaving every other surface. It would be nice to have a straightforward approximation to deal with this. The one bounce approximation can be used.

When the emissivity of all surfaces is 0.7 or greater, a simplified, approximate approach to dealing with the net radiation transfer calculations can be used. This approach also helps to visualize the transfer process for more complex geometries. Consider the radiation emitted from surface 1 in Figure 6.23. We are interested in the net exchange of radiation between surface 1 and surface 3. Some radiation emitted from 1 will go directly to 3 while another fraction of the radiation will be directed toward surfaces 2 and 4 and then be reflected from them to surface 3. If the reflectivity is 0.3, only 30% of the radiation from 1 that falls on 2 will be reflected. Part of this goes on to surface 3 and another part goes to 4. Of that latter portion, only 30% is reflected from surface 4 – just 9% of the original amount emitted from surface 1. In the succeeding two reflections, only 3% and 1%, respectively, of the original radiation is transmitted.

It follows that a reasonable approximation for high emissivity surfaces is to consider direct radiant exchange between bodies along with radiation that makes one reflection or "one bounce" from surrounding surfaces when calculating the transfer. This is illustrated in Figure 6.23. Assuming the surfaces are diffuse emitters, the amount of radiation emitted from surface 1 that is incident on surface 3 is $\varepsilon_1 \sigma T_1^4 A_1 F_{13}$ and the amount absorbed is

$$Q_{1 \rightarrow 3 \text{Direct}} = \varepsilon_1 \sigma T_1^4 A_1 F_{13} \alpha_3 \tag{6.70}$$

The amount of radiation emitted from 1 that is incident on 2 is

$$Q_{1 \rightarrow 2 \text{Incident}} = \varepsilon_1 \sigma T_1^4 A_1 F_{12} \tag{6.71}$$

and the amount from 1 to 2 that is reflected in all directions is

$$Q_{1 \rightarrow 2 \text{Reflected}} = \left(\varepsilon_1 \sigma T_1^4 A_1 F_{12} \right) \rho_2 \tag{6.72}$$

If surface 2 is a diffuse reflector with uniform incident flux over its area, the amount reflected from 2 that is directed at 3 is

$$Q_{1 \rightarrow 2 \rightarrow 3 \text{Incident}} = \left(\varepsilon_1 \sigma T_1^4 A_1 F_{12} \right) \rho_2 F_{23} \tag{6.73}$$

The amount absorbed by surface 3 becomes

$$Q_{1 \rightarrow 2 \rightarrow 3 \text{Absorbed}} = \left(\varepsilon_1 \sigma T_1^4 A_1 F_{12} \right) \rho_2 F_{23} \alpha_3 \tag{6.74}$$

By a similar argument, the amount of radiation that leaves surface 1 and is absorbed by 3 after one bounce from surface 4 can be written as

$$Q_{1 \rightarrow 4 \rightarrow 3 \text{Absorbed}} = \left(\varepsilon_1 \sigma T_1^4 A_1 F_{14} \right) \rho_4 F_{43} \alpha_3 \tag{6.75}$$

The net one-way flux from surface 1 to 3 is the sum of Eqs. (6.70), (6.74), and (6.75). This can be rewritten in terms of an overall view factor between surfaces 1 and 3 as

$$Q_{1 \rightarrow 3 \text{Total}} = A_1 \bar{F}_{13} \varepsilon_1 \alpha_3 \sigma T_1^4 \tag{6.76}$$

One can show that reciprocity holds so that $A_1 \bar{F}_{13}$ is equal to $A_3 \bar{F}_{31}$. The net radiation flux emitted by 1 absorbed by 3 less the flux emitted by 3 absorbed by 1 becomes, for gray bodies,

$$Q_{1 \rightleftarrows 3 \text{Net}} = A_1 \bar{F}_{13} \varepsilon_1 \alpha_3 \left(\sigma T_1^4 - \sigma T_3^4 \right) \tag{6.77}$$

if the emissivity and absorptivity are the same for each of the bodies.

The assumption of diffuse emitters holds reasonably well for many engineering surfaces. However, the assumptions of diffuse reflection and uniform incident intensity over a large surface area are less likely to hold. As the emissivity decreases these effects become more important. We will deal with approximations for those cases in later sections. In the following examples, we will continue to assume all five conditions mentioned after Eq. (6.67) hold.

Example 6.5 *One-bounce approximation for parallel gray surfaces* Consider a simple geometry, two parallel plane isothermal gray surfaces that are large compared to their spacing. All of the radiation from one surface is directed at the other. Of the radiation directed from surface 1 to 2, one bounce only brings that

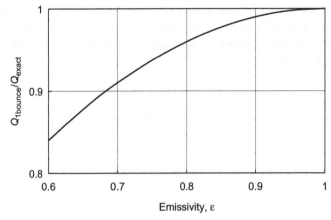

Figure 6.24. Parallel gray plates, one-bounce solution divided by exact solution as a function of emissivity.

radiation back to surface 1. Thus, in this case, only the direct component is considered in the flux from surface 1 to 2. The net radiation is approximated as

$$Q_{12\,\text{Net}} = \varepsilon_1 \varepsilon_2 A_1 F_{12}(\sigma T_1^4 - \sigma T_2^4)$$

where F_{12} is unity. Figure 6.24 illustrates that in this case for an emissivity of 0.7 or higher, the one-bounce solution is within 10 percent of the exact solution.

Example 6.6 *One-bounce approximation for an extruded rectangular object*
Similar results are obtained for other cases. Figure 6.25 shows the results for the net transfer between two parallel walls of height H within a long extruded form with a rectangular cross section with negligible transfer to the ends.

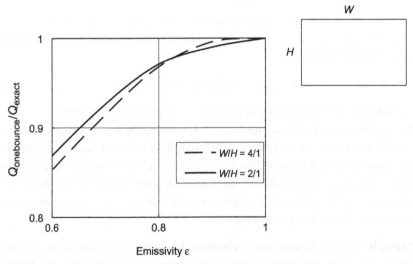

Figure 6.25. One bounce divided by exact solution for gray transfer between the end walls of a long extruded rectangle with width to height ratio, W/H, of 2:1 and 4:1 as a function of the emissivity.

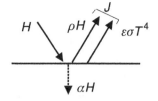

Figure 6.26. Radiative transfer to a gray surface.

For the radiation between the two walls of height H when the emissivity of all of the surfaces is 0.7 or larger the difference between the one-bounce solution and the exact is less than 10 percent. When one surface is nonblack and all surrounding surfaces are black, the one-bounce approximation holds for any value of the emissivity for the nonblack surface.

When dealing with surfaces where the diffuse approximations no longer hold a numerical method is required. For example, groups of photons emitted from one surface are followed as they interact with surrounding surfaces. In a sense this becomes a multi-bounce procedure.

General Techniques for Diffuse Gray Body Radiation

When the emissivity of one or more bodies is lower than 0.7, the one-bounce solution no longer yields good approximations. It is possible to extend the technique by including two and more reflections, but this becomes too cumbersome. A more direct method of dealing with multiple reflections is called for.

In the following development, all bodies are assumed to be gray, with a uniform emissivity over all significant wavelengths, and all bodies are treated as diffuse emitters and reflectors. The burden of these methods is the need to introduce additional definitions. Two new terms are required. The incident flux per unit surface area coming to a surface, say A_i, from all directions and in all wavelengths will be represented by H_i, called the *irradiance*. All possible sources of radiation incident on a surface, all other bodies an observer on the surface can see, must be included. Note that not all of the radiation incident on a surface is absorbed by that body. The net heat transfer per unit area to a given surface can be given for a gray body as (see Fig. 6.26).

$$q_{1net} = \varepsilon_1 \sigma T_1^4 - \alpha_1 H_1 = \varepsilon_1 \sigma T_1^4 - \varepsilon_1 H_1 \qquad (6.78)$$

The second quantity needed is the net radiation leaving a surface per unit area in all directions and all wavelengths. This will be represented by J_i, which is called the *radiosity*. The leaving energy represents the sum of the emitted plus reflected radiation from a gray body surface

$$J_1 = \varepsilon_1 \sigma T_1^4 + \rho_1 H_1 = \varepsilon_1 \sigma T_1^4 + (1 - \varepsilon_1) H_1 \qquad (6.79)$$

Subtracting Eq. (6.79) from (6.78) gives, after rearrangement,

$$q_{1Net} = J_1 - H_1 \qquad (6.80)$$

This is the energy balance an observer slightly above the surface of A_1 would make, setting the net radiation equal to the difference between the total radiation leaving a surface and the total incident radiation on that surface. Note that as the emissivity of a surface becomes very small the value of radiosity J approaches the value of the incident flux H: the surface is a good reflector and most of the incident flux is reflected while little radiation is emitted. For this case, the net radiative heat transfer for that surface approaches zero. The surface still plays a role in the overall radiant exchange, redirecting radiation from one surface to another. In contrast, for a body with emissivity near unity most of the incident flux is absorbed and the flux leaving the surface is due primarily to the radiation emitted by that surface.

If we know the temperature and emissivity of all surfaces interchanging radiation, then to find the net heat transfer to a surface, the values of the radiosities of all surfaces are needed. If all the surfaces are diffuse emitters and reflectors and all of the five assumptions given directly before Eq. (6.68) hold, the incident flux can be written as

$$Q_{1\text{Incident}} = A_1 H_1 = \sum_{k=1}^{N} A_k F_{k1} J_k = \sum_{k=1}^{N} A_1 F_{1k} J_k$$

$$H_1 = \sum_{k=1}^{N} F_{1k} J_k \tag{6.81}$$

where we have used the reciprocity relationship for the terms within the summation. The summation must be taken over all surfaces that can radiate to each other. The net heat transfer, from Eq. (6.80), is

$$Q_{1\text{Net}} = A_1 \left(J_1 - \sum_{k=1}^{N} F_{1k} J_k \right) = \sum_{k=1}^{N} A_1 F_{1k} (J_1 - J_k) \tag{6.82}$$

Although the entire summation over the N surfaces is the net radiation exchanged between surface A_1 and all of its surrounding surfaces, each term in the summation represents the net radiation leaving surface A_1 directed toward surface A_k less radiation leaving A_k directed toward A_1. Note that A_k may be highly reflecting so the term $A_1 F_{1k} (J_1 - J_k)$ does *not necessarily* represent the net radiation absorbed by A_1 that was *emitted* by A_k. Some of the energy leaving A_k may be energy from another surface reflected off A_k.

We can relate the net radiation heat transfer for one surface to its emissive power and radiosity by multiplying the emissivity by each term in Eq. (6.80) and subtracting this from Eq. (6.78). This yields

$$q_{1\text{Net}}(1 - \varepsilon_1) = \varepsilon_1 \sigma T_1^4 - \varepsilon_1 J_1 + \varepsilon_1 H_1 - \varepsilon_1 H_1 \tag{6.83}$$

Rearranging,

$$Q_{1\text{Net}} = A_1 q_{1\text{Net}} = \frac{A_1 \varepsilon_1}{(1 - \varepsilon_1)} \left(\sigma T_1^4 - J_1 \right) \tag{6.84}$$

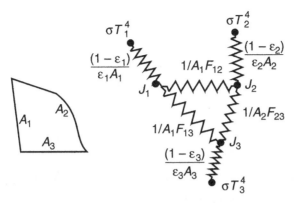

Figure 6.27. Electrical analogy for diffuse gray body radiation using radiosity.

Figure 6.27 shows the radiation network for a three body configuration, with the net flux related to the surface radiosities, Eq. (6.82), and also represented at the interconnection between the radiosity and emissive power for each surface seen in Eq. (6.84).

Example 6.7 *Nuclear fuel rod transport container with nonblack rods* Returning to the nuclear fuel rod example shown in Figure 6.19, consider the case in which the rods are not black, but are diffuse gray bodies. Further assume the rods in the each concentric row are almost touching and for simplicity the rods in rows 1 and 2 do not have any net heat generation with radiation as the sole mode of heat transfer. If we proceed as in Figure 6.22, the electric analogy that resulted would be that seen in Figure 6.28a. In that representation, however, the emissivity of the intermediate rows A_1 and A_2 does not influence the overall heat transfer. This is obviously wrong: if these two rows are good reflectors, then the net heat transfer between A_0 and A_3 should be severely reduced.

Assuming that surfaces A_1 and A_2 are reasonably isothermal and are diffuse reflectors and emitters, we need to identify another source of error. Examining the five assumptions required to use this analysis, the one that is strongly violated is the assumption that the incoming flux is uniform over the surface of each body. The circumference of rods in row 1 and 2 will see a strongly varying incident flux: the inside of row A_1 faces rod A_0 and itself, while the outside faces the rods making up row A_2.

A much better model is to break up row 1 into two separate surfaces, one that faces inward and another that faces outward. If the rods are good conductors, it is reasonable to assume that the inside and outside surfaces are at the same temperature. However, the inside and outside have very different incident fluxes and radiosities. The net radiant energy transfer received by the inside surface must equal that leaving the outside surface. This is accomplished by the circuit shown in Figure 6.28b. A similar split is made for the inside and outside surfaces of row 2. In this case the electrical analogy yields a straightforward solution of resistances in series. This circuit is in fact the same circuit that would apply to a series of concentric radiation shields.

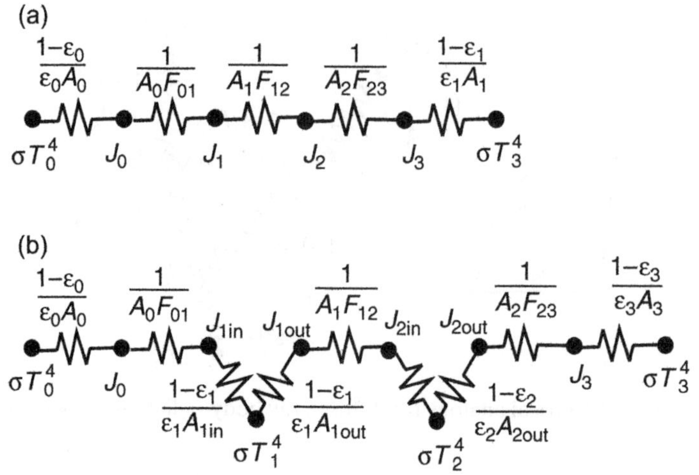

(a)

(b)

Figure 6.28. Gray fuel rod surfaces. (a) An incorrect electrical analogy. (b) A more realistic electrical analogy.

Multiple Gray Surfaces

When the assumption of uniform incident flux appears difficult to justify, each initial surface may have to be divided into multiple surfaces, each with its own incident flux and radiosity. When the number of subdivisions becomes larger than about 4, the electrical analogy usually is not much help. In this case we can write the equation for the radiosity of a subarea A_i by combining Eqs. (6.79) and (6.81):

$$J_i = \varepsilon_i \sigma T_i^4 + (1 - \varepsilon_i) H_i = \varepsilon_i \sigma T_i^4 + (1 - \varepsilon_i) \sum_{k=1}^{N} F_{ik} J_k \qquad (6.85)$$

This can be rearranged to

$$\sigma T_i^4 = \sum_{k=1}^{N} \left[\frac{\delta_{ik}}{\varepsilon_i} - \frac{1 - \varepsilon_i}{\varepsilon_i} F_{ik} \right] J_k \equiv \sum_{k=1}^{N} [B_{ik}] J_k \quad \text{where } \delta_{ik} = 1 \text{ when } i = k; \qquad (6.86)$$
$$\text{and } \delta_{ik} = 0 \text{ when } i \neq k$$

Equation (6.86) forms a series of N equations for each of the N surfaces in terms of the N unknown values of J_k. Note the quantities within the bracket defined as B_{ik} contain all of the information about surface geometry and emissivity. This series of equations can be rewritten in matrix form as

$$\bar{E} = \bar{B} \bar{J} \quad \text{and} \quad \bar{J} = \bar{B}' \bar{E} \qquad (6.87)$$

where $\bar{B}$ is the coefficient matrix and $\bar{E}$ and $\bar{J}$ are vectors containing terms for the emissive power and radiosity, respectively, of each surface. Equation (6.87) can be solved by a number of methods such as matrix inversion, to obtain the inverse matrix $\bar{B}'$ Note that no matter how the temperature distribution of the surfaces changes, if the geometry and emissivity remain constant, the matrix inversion has to be performed only once.

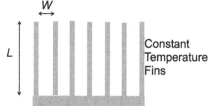

Figure 6.29. Parallel plate fins radiating to space.

Example 6.8 *Spacecraft heat sink* Figure 6.29 illustrates a large array of parallel closely spaced fins for a spacecraft heat rejection device. If the fins are at uniform temperature and are gray diffuse surfaces with an emissivity of 0.1, what is the net radiative energy transferred to space, assuming the fins are facing away from the sun and from Earth's surface?

The upper limit for this case is clearly when the radiation streaming out of the gap between the plate fins is black body radiation at the temperature of the fins. The incident flux on a fin surface will vary from the fin base to its tip. To deal with this variation, the fin surface is divided into a number of equally spaced subareas. Figure 6.30 illustrates the results for the net heat transfer as the number of subdivisions of each surface is increased. A fairly large number of subdivisions are needed to achieve reasonably accurate results. The finer divisions are needed to capture the radiosity variation near the open end of the fins. Near the middle of the fins, the incident flux on the surface will be fairly constant. Because the largest variation occurs near the open end, it would be more efficient to use close-spaced divisions near the open end and wider divisions near the fin base.

When the number of subelements becomes large, it may be more efficient to solve for the radiosities by successive approximations. That is, at the first step neglect any reflections and for each element take J_k equal to the emitted radiation from that

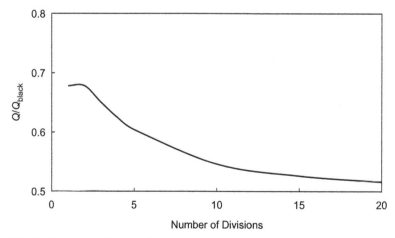

Figure 6.30. Net heat transfer from fins to space vs. number of surface divisions, surface emissivity 0.1, fin spacing to length ratio, $W/L = 10$.

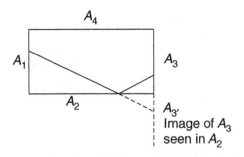

Figure 6.31. Use of mirror images with specular reflectors.

element:

$$J_k(1) = \varepsilon_k \sigma T_k^4 \tag{6.88}$$

Then, for the second guess, use this set of values of $J_k(1)$ in Eq. (6.85) for all N surfaces to get a second round of values for J_k:

$$J_i(2) = \varepsilon_i \sigma T_i^4 + (1 - \varepsilon_i) \sum_{k=1}^{N} F_{ik} J_k(1) \tag{6.89}$$

This value of $J_k(2)$ is equivalent to the results using the one bounce approximation. $J_k(2)$ is again used in the right hand side of Eq. (6.85) to get the next set of values for $J_k(3)$ and so on. This series converges rapidly when the emissivity of all the surfaces are moderate or large.

Nondiffuse Reflection

When surfaces with high reflectivity are employed, the reflection properties are more important in estimating the net radiative flux. We will deal with one limiting case, radiation exchange between bodies that include one or more specular (mirror-like) reflectors. Specular behavior will be more likely for polished metal surfaces that also have a low value of emissivity.

Consider an enclosure with four surfaces forming a rectangular cavity shown in Figure 6.31. We are interested in the net radiation between the two ends, surfaces A_1 and A_3. Along with direct radiation from A_1 to A_3, there is radiation from A_1 reaches A_3 by reflections from A_2 and A_4. When A_2 is a diffuse reflector, a substantial fraction of radiation emitted by A_1 that arrives at A_2 will be reflected back to A_1. However, when A_2 is a specular reflector, radiation from A_1 that falls on A_2 will be reflected toward A_3 (or the right end of A_4), resulting in a higher net flux from A_1 to A_3.

To approach this problem, consider the mirror image of surface A_3, labeled $A_{3'}$ that is seen by an observer on A_1 looking at surface A_2. The geometry of one ray originating from A_1 is shown, both the actual path and the path to the mirror image. The incident angle of the reflected ray on A_3 is identical to the incident angle on $A_{3'}$ as is the path length. Thus the configuration factor from A_1 to $A_{3'}$, $F_{13'}$, will represent the fraction of diffuse radiation leaving A_1 and reflected by A_2 that is directed to A_3. To be correct, this fraction must be adjusted to account for the reflectivity of A_2.

Similar to the one-bounce expression, Eq. (6.74), the amount of energy emitted by A_1 that is absorbed by A_3 by way of reflection from A_2 can be written as

$$Q_{1 \to 2 \to 3 \text{Absorbed}} = \left(\varepsilon_1 \sigma T_1^4 A_1 F_{13'} \right) \rho_2 \alpha_3 \qquad (6.90)$$

If A_4 is also a specular reflector, this construction can be extended to determine the radiation from A_1 reflected by A_2 to A_4 and then to A_3. These same kinds of expressions can be used in an electrical analogy such as Figures 6.27 and 6.28b that includes specular surfaces.

6.9 Absorbing Media

Up to now, we have assumed that radiation transferred between opaque surfaces passes through connecting volumes that are transparent. There are important instances when the volumes may absorb as well as emit radiation themselves. Examples include heat transfer in the atmosphere, in combustion chambers, in particle clouds, and in low-density thermal insulations. These media adsorb radiation over a finite thickness rather than just at their surface. They are intermediate between opaque and transparent bodies and are sometimes referred to as semitransparent bodies. We will look at the fundamental physics and establish some simplifying approaches.

Cold Media Surrounded by Hot Walls

Consider first the case of a uniform homogeneous medium such as a cloud of carbon particles, soot, where each particle can be considered a black body. The particles will be enclosed between two plane parallel walls that are black (Fig. 6.32). If the particles are cold and are surrounded by hot walls, radiation emitted by the walls will be absorbed by the particles while the radiation emitted by the particles can be neglected. To determine the amount of radiation absorbed, consider the intensity of radiation emitted from one wall at an angle θ, to its normal. As the beam of radiation intercepts particles, some of the energy will be absorbed and the intensity diminished. That change can be related to the mean free path between particles, l_{mf}, and the distance the radiation traverses, dS,

$$di(\theta) = -i(\theta) \frac{dS}{l_{\text{mf}}} \qquad (6.91)$$

and for a uniform cloud of particles, bounded by a wall that emits at an intensity i_B,

$$\frac{i(\theta)}{i_B} = e^{-S/l_{\text{mf}}} = \tau(\theta) \qquad (6.92)$$

Equation (6.91) determines the intensity of radiation emitted by the wall that has passed through a distance S without attenuation. As such, it gives the transmissivity, $\tau(\theta)$, for the angle θ; and $[1 - \tau(\theta)]$ represents the fraction absorbed by the particles.

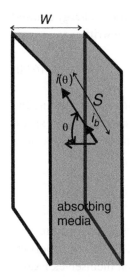

Figure 6.32. Absorbing media between plane parallel walls.

To find the total amount of energy emitted by wall A_1 absorbed by the particle cloud, we need to consider the radiation emitted by a point on the wall in all directions in a hemisphere above the surface, similar to Eq. (6.30). In this case, the integration over the hemisphere involves the distance S which varies with the angle θ as $W/\cos\theta$, where W is the spacing between the parallel plates:

$$dQ_{1\rightarrow PC} = i_{b1}dA_1 \int_{2\pi} (1 - e^{-S/l_{\text{mf}}}) \cos\theta_1 d\omega_{12} = i_{b1}dA_1 \int_{2\pi} (1 - e^{-W/l_{\text{mf}}\cos\theta_1}) \cos\theta_1 d\omega_{12} \tag{6.93}$$

Complications ensue because now the integration will not only involve geometric parameters, but also the mean free path which is a function of the density of particles. Fortunately, the integration can be approximated over the range of values for the mean free paths by the use of the *mean beam length*:

$$\int_{2\pi} (1 - e^{-W/l_{\text{mf}}\cos\theta_1}) \cos\theta_1 d\omega_{12} \approx \pi(1 - e^{-L_{\text{mean}}/l_{\text{mf}}}) \tag{6.94}$$

where the mean beam length, L_{mean}, is estimated as

$$L_{\text{mean}} \approx (0.9 \rightarrow 1.0)\frac{4V}{A_{\text{surface}}} \tag{6.95}$$

where V is the volume of the particle cloud and A_{surface} is the total area of the surfaces enclosing the particles. In the case of parallel plates, $4V/A_{\text{surface}}$ is just twice the separation distance, W, when the plate dimensions are large compared to W. Thus, the heat flux from one wall to the particles can be approximated as

$$dQ_{1-PC} = dA_1\sigma T_1^4(1 - e^{-L_{\text{mean}}/l_{\text{mf}}}) = dA_1\sigma T_1^4(1 - e^{-2W/l_{\text{mf}}}) \tag{6.96}$$

Uniform Temperature Hot Medium

When the particles are hot and well mixed, so that their temperature is uniform, they will emit significant radiation to the wall as well as absorb radiation from the wall. For a very dilute mixture of particles, an upper limit for the radiation to the bounding walls from the particles is found by assuming that the radiation from each particle reaches the wall:

$$Q_{PC-\text{walls}}(\text{MAX}) = \sigma T_{PC}^4 \sum A_{PC} \tag{6.97}$$

This estimate will overpredict the radiant flux to the walls because some of the radiation emitted by individual particles will be absorbed by neighboring particles before it reaches the walls. If the temperature of the particles is the same as the wall temperature, then the amount of radiation from the particles to a wall must be just equal to the radiation from the wall that is absorbed by the particles given by Eq. (6.96).

Consider a more accurate upper limit of particle radiation to the surrounding walls. When the particle cloud is at a temperature T_{PC}, the radiation emitted by the particles to a subsection dA_1 of the wall, over all incident angles θ_1, becomes

$$dQ_{PC-dA1} = i_{bPC}dA_1 \int_{2\pi} \left(1 - e^{-W/l_{mf}\cos\theta_1}\right) \cos\theta_1 d\omega_{12} = dA_1 \sigma T_{PC}^4 (1 - e^{-L_{mean}/l_{mf}}) \tag{6.98}$$

In the limit, when L_{mean} is much larger than the mean free path l_{mf}, either as a result of large plate spacing W or high particle density, the cloud of particles emits as much radiation as an opaque black body. At the same time, it absorbs all of the incident radiation emitted by the bounding wall: it has an absorptivity of unity.

Gas Radiation

Many common gases absorb radiation as incident photons excite a transition between different energy states of the gas molecule. Gases do not absorb uniformly over all wavelengths; rather, they tend to absorb in distinct ranges of wavelengths or *bands*. This gives rise to the so-called greenhouse effect for the Earth's atmosphere because gases such as carbon dioxide are transparent over the short solar wavelengths but have absorption bands at the longer wavelengths where infrared radiation is emitted by the earth's surface. The amount of gas absorption increases with the gas density, so rising CO_2 concentrations can increase the greenhouse effect.

As we have seen with the particle cloud, if a medium absorbs more radiation, then it will also emit more. In addition, the wavelength span and magnitude of the absorption band is influenced by the gas pressure and temperature as well as the density of other gases in the mixture. Experimental measurements and theoretical models have been used to integrate these effects so as to produce a measure of gas emission over all wavelengths for a well-mixed gas at uniform temperature. One such result, shown in Figure 6.33, is for carbon dioxide in terms of the effective emissivity

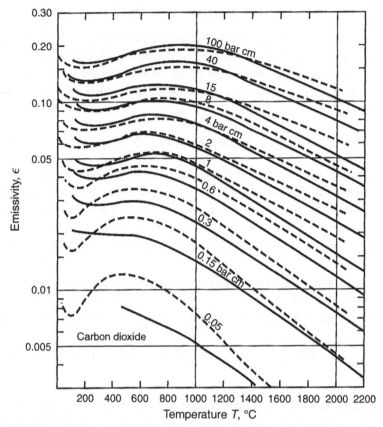

Figure 6.33. Emissivity of CO_2 as a function of a path length L, times partial pressure p_{CO2} in a gas mixture with a total pressure of one bar. *Sources:* Hottel [6.3] solid line; Leckner [6.4] dashed line.[1]

of the gas. The effective emissivity is the ratio of the radiation emitted by the gas to a bounding surface area dA_1 from all directions to the radiation a black body at the gas temperature would emit to dA_1:

$$\varepsilon_g = \frac{dQ_{g-dA_1}/dA_1}{\sigma T_g^4} \tag{6.99}$$

The dashed lines represent approximate extrapolations from the measured data. For a uniform temperature gas of arbitrary shape, the mean beam length, Eq. (6.95), can be used to evaluate the path length L in the figure.

Because the gas does not absorb uniformly over all wavelengths, a correction to the results must be used to calculate the absorptivity of the gas to radiation from a source at temperature substantially different than the temperature of the gas.

---

[1] Because the pressure level influences pressure broadening of absorption lines and bands, the emissivity is a not solely a function of the $p_{CO2}L$ product but is also a modest additional function of the partial pressure and the total pressure of the mixture. Hottel and Leckner have used an adjusted value of the emissivity to allow these two pressure corrections to be dealt with in a simple form. At modest pressures, one bar or less, these corrections are only a few percent [6.4].

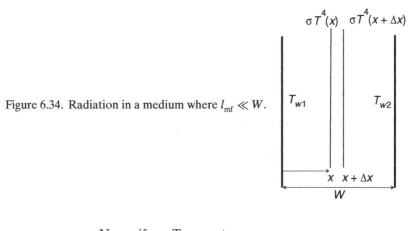

Figure 6.34. Radiation in a medium where $l_{mf} \ll W$.

Nonuniform Temperatures

When there are noticeable temperature gradients across an absorbing medium, the concept of effective emissivity loses most of its utility. An upper and lower limit to the radiation emitted from the medium can be obtained by assuming a well-mixed medium at either extreme temperature. In some instances, however, the temperature distribution of the medium is initially unknown. One numerical method of solution divides the medium into smaller zones and the exchange between individual zones is accounted for [6.5]. Other analytic and approximate methods are given, for example, by Modest [6.6] and by Howell and Siegel [6.7].

One limiting case becomes straightforward to deal with. When the medium absorbs over all wavelengths that have significant blackbody radiation and the mean free path for every wavelength is much smaller than the linear dimensions of the enclosure, a diffusion-like formulation can be obtained. This can be illustrated for the case of a rigid medium contained between two plane parallel walls that are at different temperatures. A one-dimensional temperature field with isothermal planes is set up across the medium as shown in Figure 6.34.

Because the mean free path, l_{mf}, is small, radiation emitted by one subvolume is absorbed over a distance of roughly l_{mf}. This approximates the radiative transfer between two black planes separated by a distance $\Delta x = l_{mf}$. Then the radiative flux is

$$q_{rad\,net} = \sigma[T^4(x + \Delta x) - T^4(x)] \approx l_{mf}\frac{d(\sigma T^4)}{dx} \qquad (6.100)$$

A more rigorous derivation shows that the proper interval is $4/3\,l_{mf}$. The final expression is usually written in terms of a mean extinction coefficient, κ, which is the inverse of l_{mf}. The mean extinction coefficient is averaged over all wavelengths containing appreciable black body radiation. The final expression, known as the Rosseland equation, is

$$q_{rad} = -\frac{4}{3\kappa}\frac{d(\sigma T^4)}{dx} = -\frac{16}{3\kappa}\sigma T^3\frac{dT}{dx} \qquad (6.101)$$

Similar to the Fourier equation, in multidimensional cases the radiative transfer rate can be written as the gradient of σT^4. In the linearized form, the radiation term can be added to the conduction term to find the overall heat transfer in the absence of convection. One caution: the mean free path must be much smaller than the linear dimensions of the enclosure for every wavelength where there is appreciable energy in the black body spectrum evaluated at the temperature level in question. For a typical gas, the wavelength gaps between the absorption bands have a high mean free path or will approach the transparent limit and will yield an erroneously high value for the total radiation transfer if Eq. (6.102) is used for all wavelengths. Instead, the equation only applies to the bands that have a mean free path much less than the physical dimensions of the gas cloud.

When the Rosseland equation applies over the entire spectrum for a medium contained between two parallel black walls with a spacing W, the net radiation heat transfer at steady state becomes

$$Q_{\text{rad}} = -\frac{4A}{3\kappa}\frac{\Delta(\sigma T^4)}{W} = \frac{4A}{3\kappa}\frac{\left(\sigma T_{w1}^4 - \sigma T_{w2}^4\right)}{W} \tag{6.102}$$

Note that very little of the radiation emitted by one bounding wall arrives at the other wall for a small mean free path because the transmissivity decreases exponentially with l_{mf} or $1/\kappa$. However, the net radiation only varies linearly with l_{mf} so that this quantity may well be an important element of heat transfer.

REFERENCES

[6.1] L. Hanssen, B. Wilthan, J.R. Filtz, J. Hameury, F. Girard, M. Battuello, J. Ishii, J. Hollandt and C. Monte, Infrared spectral normal emittance/emissivity comparison, *Metrologia*, **53**(1A), 03001, *Technical Supplement*, 2016.

[6.2] J. Howell. *A catalog of radiation heat transfer configuration factors.* http://www.thermalradiation.net/tablecon.html

[6.3] H.C. Hottel. Radiant heat transmission. In W. H. McAdams (ed.), *Heat Transmission*, 3rd ed. New York: McGraw-Hill, 1954, Chapter 4.

[6.4] B. Leckner. Spectral and total emissivity of water vapor and carbon dioxide. *Combustion and Flame*, **19**(1), 33–48, 1972.

[6.5] H.C. Hottel and A. F. Sarofim. *Radiative Transfer.* New York: McGraw-Hill, 1967.

[6.6] M. Modest. *Radiant Heat Transfer*, 3rd ed. New York: McGraw-Hill, 2013.

[6.7] J. Howell and R. Siegel. *Thermal Radiation Heat Transfer*, 5th ed. New York: McGraw-Hill, 2010.

PROBLEMS

6.1 For a nonblack surface at 1000 K how would you distinguish between reflected solar energy and emitted energy leaving the surface?

To measure the emissivity, it is proposed to put a sample at the center of a tube furnace kept at 1000 K and measure the radiant energy from the surface to the open end of the tube furnace. The open end is at room temperature. How well will this work?

To determine the reflectivity of the 1000 K surface an inventor proposes to use a source at 1200 K. How would you distinguish between emitted and reflected energy in this case?

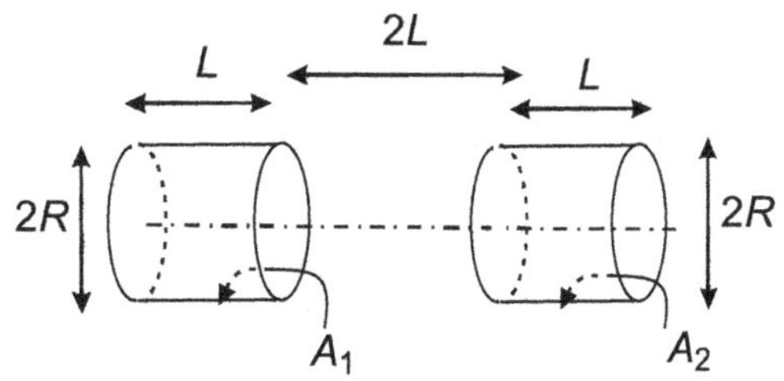

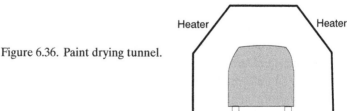

Figure 6.35. Sections of tube furnace.

6.2 A tube furnace is a simple means to achieve high temperature (Fig. 6.35). It is a cylindrical ceramic tube wrapped with electrical resistance wire. Due to end losses the center is much hotter than the ends. To calculate the center temperature we need to know the configuration factors between various internal sections of the empty tubes. Find the configuration factors between the two sections shown. Try to use the shape factors in the appendix and shape factor algebra.

6.3 You are asked to develop an appropriate technique to estimate the configuration factor between two right circular cylinders of radius R and length L based on the known configuration factors given, for example, online. The axes of the two cylinders are parallel and the bases are in the same plane. The axes of the two cylinders are separated by a distance of S. We are interested in cases where $S > 5R$ and $L \geq 2R$.

Base your approximation on the known configuration factor between two surfaces of finite length given in the online catalogue of configuration factors. Check by comparing the approximation to the exact solution for infinitely long parallel cylinders.

6.4 A long "tunnel" is used for paint drying in the automotive industry (Fig. 6.36). Minivans move in a continuous stream next to each other so they look like a continuous obstacle running the length of the tunnel. Take the three heated surfaces as black bodies at the same temperature. Show graphically how you would determine the percentage of radiation emitted from the heaters that is directed to the perimeter of the minivan, and that directed from the heaters to the floor. Does this add up to 100%?

Figure 6.36. Paint drying tunnel.

6.5 We can use the techniques developed for thermal radiation to determine the distribution and reflection of visible light (and vice versa). Consider diffuse light entering a window with uniform intensity in all directions. Immediately in front of the window are horizontal blinds of width L with a spacing of w between successive blinds. The length of the blind Z is much larger than L or w. The bottom of the blinds is painted black. The top is a mirror-like surface with reflectivity 0.9. Find the fraction of light entering the window between the two blinds that reaches the ceiling directly and the fraction reflected off the blind surface that reaches the ceiling. For the latter, consider the image of the window area $w \cdot Z$ seen in the mirror surface by an observer on the ceiling. The ceiling has dimensions $X \cdot Z$ and it is Y distance above the upper blind (Fig. 6.37). Take $w = 4$ cm, $L = 5$ cm, $Y = 10$ cm, and $X = 30$ cm.

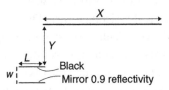

Figure 6.37. Reflective window blinds.

6.6 A plane metal ceramic surface is heated to a uniform temperature T_o of 1000°F to act as a radiant heater. The ceramic has an emissivity of 0.5 and is a diffuse emitter and reflector. A designer proposed to groove the surface as shown to increase its effective emissivity ε_a (Fig. 6.38). Here, ε_a is defined in terms of the imaginary surface A_s as $\varepsilon_a = (Q/A_s)/\sigma T_o^4$ where Q is the flux passing out through A_s.

  a. By dividing one V-groove cavity into two isothermal elements, the two sides, find an expression for ε_a versus 2ϕ from 0° to 180°.

  b. Is the limit at $2\phi \to 0°$ correct? Consider the diffuse surface cavity as an absorber of incident radiation. Does $\alpha \to 1.0$ as $2\phi \to 0°$? Do calculations to support your conclusions.

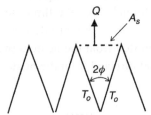

Figure 6.38. Effective emissivity of grooved surface.

6.7 Figure 6.39 shows a cross section of a typical two-by-four framed building. One insulation technique is the use of reflective foil, which is placed over the inside of one sheet of wallboard (sometimes it is installed halfway between the two sheets of wallboard). The top wall is at a uniform temperature of T_1, the bottom at a uniform temperature T_2. Assume that the wall section is horizontal with $T_1 > T_2$ so that the air is stationary. Experimental results for the cross section were substantially higher than predictions using a one-dimensional model for the air space and stud.

  a. Calculate the net heat transfer per unit cross-sectional area from T_1 to T_2 assuming one-dimensional heat transfer through the air space and wood stud.

b. Find the heat transfer through the wood from T_1 to T_2. Include conduction through the wood as well as radiant exchange from surfaces A_1 and A_2 to the side of the wood. Develop an approximate method to model the net radiant transfer from A_1 and A_2 to an element of thickness dx in terms of the local wood temperature $T(x), T_1, T_2,$ and the emissivities. Try to simplify with engineering approximations. You should be able to approximate the net radiant flux to an element on the side of the wood as $h_r(T_1 - T_x)dx$ where x is the coordinate along the 3½ inch length. Include the radiant term in the standard fin equation for the wood and develop a closed form solution. How much does this increase the heat transfer from T_1 to T_2?

Neglect air conduction or convection to the stud. Note that two-by-fours have actual dimensions of 3/2 in. by 3 1/2 in. and that they are on a center-to-center spacing of 16 in.

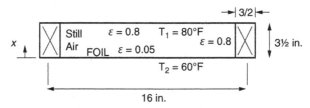

Figure 6.39. Wall section with reflective insulation.

6.8 The fins of a plate fin exchanger form a series of very long plane parallel plates (Fig. 6.40). Consider a spacing-to-width ratio of 0.05. The plates are gray diffuse surfaces with $\varepsilon = 0.1$. For the simplest case, assume that they are at the same uniform temperature. Radiation from the environment to the plates can be neglected.

Using the lumped system approximation and symmetry conditions to simplify the problem as much as possible, calculate the net heat transfer from one surface if each surface is divided into three elements as shown.

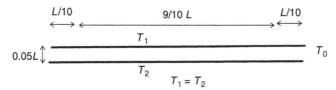

Figure 6.40. Radiation between fins.

6.9 To reduce the radiation heat transfer to a wall surrounding a furnace an inventor suggests attaching low-conductivity spheres to the wall (Fig. 6.41). The spheres are uniformly spaced in a square array with a center-to-center spacing of S. The gap between the two walls, W, is much smaller than the length of the walls. Assume each surface can be approximated as an isothermal black body. The furnace is at 1100 K and the surrounding wall is at 400 K. Estimate the direct radiant heat transfer per unit area between the furnace wall and the surrounding wall. Do not include the radiation emitted by the spheres. Do your estimate for two cases:

a. The center-to-center spacing of the spheres, S/D, is 5. Note that for this case it is not adequate to assume that the spheres are absent.

b. The center-to-center spacing of spheres, S/D, is 1. They are touching. For this case only make an upper limit estimate for the direct radiant heat transfer.

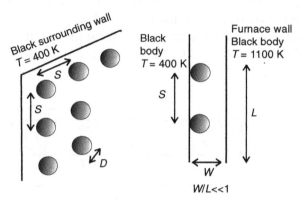

Figure 6.41. Radiation shield using spheres.

6.10 a. Gas is flowing through a long 200 mm pipe. The inside pipe surface can be assumed to be a diffuse surface with an emissivity of 0.3. The pipe wall has a uniform temperature of 200°C. Hot gas enters the pipe, and at the location in question, far from the entrance, the gas temperature is 300°C. A 1 mm diameter spherical thermistor is used to measure the gas temperature. The thermistor is supported by a solid 5 mm diameter stainless rod that extends 20 mm from the pipe wall. The rod has a thermal conductivity of 10 W/m K, and its surface can be assumed to be black as a limiting case. The convective heat transfer coefficient from the gas to the rod is 10 W/m²K. The convective heat transfer on the thermistor is 25 W/m²K. At steady state, estimate the temperature that the thermistor records.

b. In an effort to reduce the error due to radiation, a cylindrical radiation shield is positioned concentric with the pipe. The shield is made of thin aluminum that has a rough surface with an emissivity of 0.1. The shield has a diameter of 25 mm and is 25 mm long. The convection heat transfer on its inside and outside surface is 10 W/m²K. Now the 1 mm diameter spherical thermistor is supported by very thin wires at the center of the radiation shield. Estimate the temperature recorded by the thermistor. It will help to first estimate the temperature of the shield.

6.11 To measure the radiation characteristics of a closed-cell foam insulation, a thin 2.5 mm thick sample is used. For a path normal to the foam surface, the transmissivity is found to be 0.13, and it remains approximately constant over all wavelengths of interest. The effective conductivity of a 2.5 cm foam sample is found to be 0.030 W/m K at room temperature.

a. What is the transmissivity for the 2.5-cm sample?
b. Is radiation heat transfer negligible for the 2.5-cm sample?
c. If not, estimate its contribution to the effective conductivity.

Additional Notes on Internal Flow Heat Transfer

A.1 Power Law Correlations Compared to Gnielinski's Equation

It is of interest to compare the predictions of the Gnielinski equation, (4.19), to the older power law correlations, such as the Dittus-Boelter equation[1]:

$$\text{Nu}_D = 0.0243 \, \text{Re}_D^{0.8} \text{Pr}^{0.4} \tag{A.1}$$

Figure A.1 shows the error in the Dittus-Boelter equation, measured against the Gnielinski equation, as a function of Reynolds number for $\text{Pr} = 0.7$, 10, and 100. The Dittus-Boelter equation shows extreme errors in the range $\text{Re}_D \leq 10^4$, but even at higher Reynolds numbers it is clearly unreliable.

Part of the difficulty with the Dittus-Boelter equation can be overcome by lowering the lead coefficient (from 0.0243 to 0.023, another commonly used value), which improves accuracy in the Prandtl number range of gases; however, the dependence on Reynolds number is not the same for higher Prandtl numbers.

The suggestion to use power law fits can be traced to Nusselt in 1909, but a year later Prandtl proposed the basic form of the Gnielinski equation, which is based upon a physical model of a two-layer boundary layer structure (for details, see the notes in [A.3]).

A.2 The Choice of Temperature and Pressure as Independent Variables

For a pure substance, two thermodynamic properties are needed to fix the state. In general, temperature is the most convenient choice for one of these variables in the study of heat transfer. First, temperature is the potential that drives heat flow according to Fourier's law. Second, temperature is easily measurable, whereas alternative thermodynamic quantities, such as entropy, are not. Likewise, pressure is often preferred as the second thermodynamic variable for pure substances because it is also

---

[1] Several similar equations are in the literature, differing mainly in the value of the lead constant: 0.021, 0.023, and 0.024 have all been suggested. The names McAdams and Colburn are also applied to this kind of equation. Dittus and Boelter recommended the version cited here for heating a fluid; they proposed a slightly different version for cooling. Further, the history of attribution for these equations is convoluted, with apparent exchange of ideas between McAdams and Dittus and Boelter [A.1, A.2].

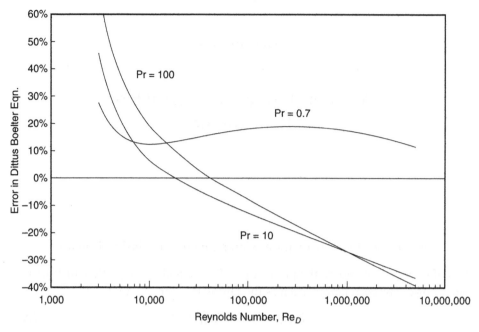

Figure A.1. Error in the Dittus-Boelter equation as a function of Reynolds number for various Prandtl numbers.

easily measured and is a primary factor in associated mechanical processes, such as fluid flow.

Temperature and pressure cannot always be used as independent thermodynamic variables to characterize the state of a pure substance. For example, in a two-phase situation T and p are not independent, so another variable is required (e.g., the quality). In contrast, other pairs of variables will allow a full characterization of the thermodynamic state. For instance, $h = h(p, s)$ gives a thermodynamic *fundamental equation* from which all other thermodynamic properties may be obtained.

One may also wonder about the use of enthalpy in our formulation, rather than internal energy or the Helmholtz or Gibbs energies. The reason for this choice is that enthalpy has a natural association with heat addition at constant pressure whereas the other three potentials naturally relate to processes at, respectively, constant volume, constant temperature, or constant temperature and pressure (see [A.4, Chapter 6]). As previously noted, within any cross section of a duct, pressure is essentially uniform while temperature (and perhaps density) may vary. The impact of the axial variation in pressure appears as the second term of Eq. (4.38).

A.3 Dissipation of Mechanical Energy

We begin with a thermodynamic identity, $dh - dp/\rho = T\,ds$, for s the entropy. Casting this into a one-dimensional bulk flow model gives

$$\frac{dh_b}{dx} - \frac{1}{\rho}\frac{dp}{dx} = T_b\frac{ds_b}{dx} \tag{A.2}$$

where ρ and p are assumed to be uniform in any cross section, and s_b is an appropriately defined bulk value. We set $\dot{V} = \dot{m}/\rho$ and use the energy equation, Eq. (4.32), to eliminate the enthalpy term:

$$T_b \frac{ds_b}{dx} = q_w \mathcal{P} + \left(\dot{W}_l - \frac{\dot{m}}{2} \frac{d}{dx} \overline{u^2} - \dot{m} \frac{1}{\rho} \frac{dp}{dx} \right) \tag{A.3}$$

The first right-hand side term represents the entropy increase due to heat transfer; the second represents entropy increase due to the difference between work input, acceleration of the flow, and flow work: it is the mechanical energy that is irreversibly dissipated, for example by friction or as electrical work resistively dissipated. Thus, we identify the mechanical energy dissipation as

$$\dot{Q}_{\text{diss}} = \dot{W}_l - \frac{\dot{m}}{2} \frac{d}{dx} \overline{u^2} - \dot{V} \frac{dp}{dx} \tag{A.4}$$

A.4 Bulk Temperature Variation with Significant Dissipative Heating

Liquids with non-negligible dissipation in the incompressible approximation. We may use the incompressible energy equation, Eq. (4.49), as a starting point.

For a given heat flux distribution, $q_w(x)$, the bulk temperature variation is

$$T_b(x) = T_{b\text{in}} + \int_0^L \frac{\mathcal{P}}{\dot{m} c_p} (q_w(x) + \dot{Q}_{\text{diss}}) \, dx \tag{A.5a}$$

For a constant flux and constant $\dot{Q}_{\text{diss}}$, this is simply

$$T_b(x) = T_{b\text{in}} + \frac{\mathcal{P}}{\dot{m} c_p} (q_w + \dot{Q}_{\text{diss}})x \tag{A.5b}$$

which is a straight line variation of T_b. The dissipation rate will be constant if the flow is fully developed with constant or zero $\dot{W}_l$ and negligible kinetic energy change.

For a uniform wall temperature passage, we must account for an additional consequence of dissipation – its effect on the temperature distribution. Dissipation is greatest where the velocity gradients are highest, and so most of the dissipation occurs near the walls. As a result, the heat transfer rate cannot be predicted accurately using the wall-to-bulk temperature difference unless new expressions for the heat transfer coefficient are developed for each duct flow. Although some such results are available (see [A.5]), they lead to oddly varying heat transfer coefficients (which may even become negative!?). Fortunately, a better approach is possible using the superposition principle.

For flows in which the velocity field is not affected by temperature, one can superpose the solution for an adiabatic walled duct with dissipation upon the usual non-dissipative solution with heat transfer at the wall. The details of this superposition calculation are straightforward (see, e.g., [A.6]), and the result is that wall heat flux may be computed as

$$q_w = h \left(T_w - T_{\text{aw}} \right) \tag{A.6}$$

in which h is the ordinary heat transfer coefficient that one would compute in the absence of dissipation, T_w is the wall temperature, and T_{aw} is the adiabatic wall temperature – the one obtained when dissipation occurs in an adiabatic passage.

The precise expression for the adiabatic wall temperature must be determined on a case-by-case basis. For fully developed laminar internal flow, typically,

$$T_{aw} = T_b + C\left(\frac{\mu u_b^2}{k}\right) \tag{A.7}$$

For flow in a circular tube, $C = 1$, and for flow between infinite parallel plates, $C = 27/35$. For turbulent internal flow, less information is available.[2]

If we introduce the inlet temperature difference, $|T_w - T_{bin}|$, Eq. (A.7) may be written in terms of the Brinkman number [Eq. (4.48)]

$$T_{aw} = T_b + C\,|T_w - T_{bin}|\,Br \tag{A.8}$$

which shows that a small Brinkman number ensures both that $\dot{Q}_{diss}$ is small and that $T_{aw} \approx T_b$.

The difference between T_{aw} and T_b is likely to be important only when μ is large, as for oils. For fluids such as air and water, dissipation matters only at the high bulk velocities. On the other hand, the influence of $\dot{Q}_{diss}$ on bulk temperature may be important even for lower speeds and viscosities if the duct is long and adiabatic.

Upon substituting Eq. (A.6) into Eq. (4.49), we have

$$\dot{m}c_p\frac{dT_b}{dx} = h(x)\mathcal{P}\left[T_w(x) - T_{aw}(x)\right] + \dot{Q}_{diss} \tag{A.9}$$

For fully developed laminar internal flow, Eq. (A.7) will apply and $\dot{Q}_{diss}$ will be constant; if we additionally set T_w constant, our governing equation is

$$\dot{m}c_p\frac{dT_b}{dx} = h\mathcal{P}\left[T_w - T_b(x)\right] + \left[\dot{Q}_{diss} - h\mathcal{P}C\left(\frac{\mu u_b^2}{k}\right)\right] \tag{A.10}$$

and it has the solution

$$T_b(x) = \left[T_w + \frac{\dot{Q}_{diss}}{h\mathcal{P}} - C\left(\frac{\mu u_b^2}{k}\right)\right] + \left[T_{bin} - T_w - \frac{\dot{Q}_{diss}}{h\mathcal{P}} + C\left(\frac{\mu u_b^2}{k}\right)\right]\exp\left(-\frac{h\mathcal{P}}{\dot{m}c_p}x\right) \tag{A.11}$$

Ideal gas flow at low speed with non-negligible dissipation and no work transfer. In this case, Eq. (4.51) reduces to Eq. (4.58). For a given wall heat flux, the solution is again given by Eq. (4.59a) or Eq. (4.59b). For a specified wall temperature distribution, Eq. (4.58) still applies, but the determination of h more complex. Flow work and dissipation both appear in the convective energy equation that is used to calculate h. These effects are of the same order in an ideal gas, so that flow work cannot be neglected if dissipation is significant, despite the fact that the two effects cancel from

---

[2] Measurements by McAdams et al. [A.7] in high velocity air gave $T_{aw} - T_b = 0.88(u_b^2/2c_p)$, implying that $C = 0.61$, but this cannot be taken as a general result, especially for other values of the Prandtl number.

the bulk thermal energy equation. The upshot is that a different adiabatic wall temperature will be required for gases than the incompressible liquid result, Eq. (A.7). This problem has been studied in the literature from time to time (see [A.5, pp. 183–184] and [A.8]).

The result so obtained is rather specialized anyway, in the sense that rather high speeds are required for dissipation to be of any significance. Specifically, consider the value of $\mu u_b^2/k$, which gives the order of magnitude of dissipation temperature rise. For air at $u_b = 10$ m/s, it is 0.07 K; at 50 m/s, it is 1.8 K; and at 100 m/s, it is 7.0 K. For speeds at which this term is large enough to matter, density variations in the streamwise direction will be significant unless the duct is relatively short. Hence, the neglect of streamwise kinetic energy changes may not be justified, and analysis should be based on the total energy equation using the methods of compressible flow theory.[3]

REFERENCES

[A.1] R. H. S. Winterton. Where did the Dittus and Boelter equation come from? *International Journal of Heat and Mass Transfer*, **41**(4–5), 809–810, 1998.

[A.2] W. C. Williams. If the Dittus and Boelter equation is really the McAdams equation, then should not the McAdams equation really be the Koo equation? *International Journal of Heat and Mass Transfer*, **54**, 1682–1683, 2011.

[A.3] R. K. Shah and M. S. Bhatti. *Turbulent Convective Heat Transfer in Ducts*. In S. Kakaç, R. K. Shah, and W. Aung (eds.), *Handbook of Single-Phase Convective Heat Transfer*. New York: Wiley-Interscience, 1987, Chapter 4.

[A.4] H. B. Callen. *Thermodynamics and an Introduction to Thermostatistics*, 2nd ed. New York: John Wiley & Sons, 1985.

[A.5] R. K. Shah and A. L. London. *Laminar Flow Forced Convection in Ducts*. New York: Academic Press, 1978 (Supplement 1 to the series *Advances in Heat Transfer*).

[A.6] P. H. Oosthuizen and D. Naylor. *An Introduction to Convective Heat Transfer Analysis*. New York: McGraw-Hill, 1999, Section 3.7.

[A.7] W. H. McAdams, L. A. Nicolai, and J. H. Keenan. Measurements of recovery factor and coefficients of heat transfer in a tube for subsonic flow of air. *Transactions of AIChE*, **42**, 907–925, 1946.

[A.8] J. Madejski. Temperature distribution in channel flow with friction. *International Journal of Heat and Mass Transfer*, **6**, 49–51, 1963.

---

[3] For high speed boundary layers in gases, well-known results show that dissipation scales with the freestream Mach number, which also scales the freestream compressibility. These two effects are different, and separate, but they are often conflated in the literature on boundary layer theory. For internal flows, the bulk conditions can be characterized by a Mach number, but other factors such as duct length, heat addition, and so on will determine whether the bulk density varies enough in the streamwise direction to affect the kinetic energy in the total or mechanical energy equations. We therefore avoid the obvious temptation to write the dissipation term for gases in the form of a Mach number squared or, equivalently, an Eckert number.

Configuration Factors Between Two Bodies of Finite Size

This appendix provides an abbreviated set of relationships for configuration factors between two bodies of finite size [6.2].

1. Two finite rectangles of the same length, having one common edge, at an angle of 90° to each other.

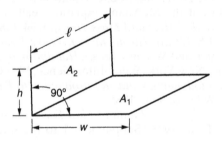

Definitions: $H = h/1 \quad W = w/1$

Governing equation:

$$F_{1-2} = \frac{1}{W\pi} \left(W \tan^{-1} \frac{1}{W} + H \tan^{-1} \frac{1}{H} - \sqrt{H^2 + W^2} \tan^{-1} \sqrt{\frac{1}{H^2 + W^2}} \right.$$

$$\left. + \frac{1}{4} \ln \left\{ \frac{(1+W^2)(1+H^2)}{1+W^2+H^2} \left[\frac{W^2(1+W^2+H^2)}{(1+W^2)(W^2+H^2)} \right]^{W^2} \left[\frac{H^2(1+H^2+W^2)}{(1+H^2)(H^2+W^2)} \right]^{H^2} \right\} \right)$$

2. Identical, parallel, directly opposed rectangles:

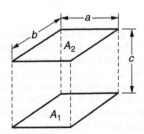

Definitions: $X = a/c \quad Y = b/c$

Governing equation:

$$F_{1-2} = \frac{2}{\pi XY} \left\{ \begin{array}{l} \ln \left[\dfrac{(1+X^2)(1+Y^2)}{1+X^2+Y^2} \right]^{1/2} + X\sqrt{1+Y^2} \tan^{-1} \dfrac{X}{\sqrt{1+Y^2}} \\[2ex] + Y\sqrt{1+X^2} \tan^{-1} \dfrac{Y}{\sqrt{1+X^2}} - X\tan^{-1}X - Y\tan^{-1}Y \end{array} \right\}$$

3. Two ring elements on the interior of a right circular cylinder:

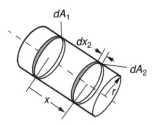

Definitions: $X = x/2r$

Governing equation:

$$dF_{d1-d2} = \left[1 - \frac{2X^3 + 3X}{2(X^2+1)^{3/2}} \right] dX_2$$

4. Disk to parallel coaxial disk:

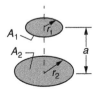

Definitions: $R = r/a; \ X = 1 + (1 + R_2^2)/R_1^2$

Governing equation:

$$F_{1-2} = \frac{1}{2} \left\{ X - \left[X^2 - 4\left(\frac{R_2}{R_1}\right)^2 \right]^{\frac{1}{2}} \right\}$$

Index